Additional Skill and Drill Manual

Cathy Ferrer
Valencia Community College

College Algebra

Tenth Edition

Margaret L. Lial
American River College

John Hornsby
University of New Orleans

David I. Schneider
University of Maryland

Boston San Francisco New York
London Toronto Sydney Tokyo Singapore Madrid
Mexico City Munich Paris Cape Town Hong Kong Montreal

Reproduced by Pearson Addison-Wesley from electronic files supplied by the author.

Publishing as Pearson Addison-Wesley, 75 Arlington Street, Boston, MA 02116.

ISBN-13: 978-0-321-53048-6
ISBN-10: 0-321-53048-9

1 2 3 4 5 6 BB 10 09 08 07

Contents

Chapter R Review of Basic Concepts

Chapter 1 Equations and Inequalities

Chapter 2 Graphs and Functions

Chapter 3 Polynomial and Rational Functions

Chapter 4 Exponential and Logarithmic Functions

Use set notation, and list all the elements of each set.

1. $\{20,21,22,...,29\}$

2. $\{4,8,12,...,28\}$

3. $\frac{1}{3},\frac{1}{9},\frac{1}{27},...,\frac{1}{729}$

4. $\{51,48,45,...39\}$

Identify the following sets as finite or infinite.

5. $\{9,18,27,...\}$

6. $1,\frac{1}{3},\frac{1}{9},...$

7. $\{x|x \text{ is a natural number less than } 6\}$

8. $\{x|x \text{ is a rational number}\}$

Complete the blanks with either $\in$ or $\notin$ so that the resulting statement is true.

9. 9 _____ $\{3,6,9,12,15,18\}$

10. $\varnothing$ _____ $\{\varnothing\}$

11. -2 _____ $\{x|x \text{is a natural number less than } 6\}$

12. $\{4\}$ _____ $\{0,2,4,8,16\}$

Tell whether each statement is true or false.

13. $19 \notin \{14,16,18,...\}$

14. $\{5,19,45,4\}=\{4,5,9,45\}$

15. $\{2,4,6\}\cap\{1,3,5\}=\varnothing$

16. $A\cup\varnothing=A$

Let $U=\{1,2,3,4,...,20\}$, $A=\{2,4,6,...20\}$, $B=\{1,5,9,13,17\}$, $C=\{3,6,9,12,15,18\}$. Tell whether each statement is true or false.

17. $C\subseteq A$

18. $\varnothing\subseteq C$

19. $B\subseteq U$

20. $A\subseteq B$

Let $U=\{1,2,3,4,...,10\}$, $A=\{2,4,6,...10\}$, $B=\{1,3,5,...,9\}$, $C=\{3,6,9\}$. Use these sets to find each of the following.

21. $A\cap B$

22. $A\cup B$

23. $B\cap C$

24. $A\cup C$

Let set $A = \left\{-3, \frac{2}{3}, \frac{7}{0}, \pi, \sqrt{13}, -\frac{8}{2}, \frac{1}{9}, 5\right\}$. List all the elements of A that belong to each set.

1. Natural numbers
2. Integers
3. Rational numbers
4. Irrational numbers

Evaluate each expression.

5. $(-2)^3$
6. -3^2
7. $-3 \cdot 2^4$
8. 5^3

Evaluate each expression.

9. $3^2 - 8 \div 4$
10. $(9+3) \div 4 \cdot 2 + 6$
11. $5 - 2(4^2 - 7)$
12. $\dfrac{-(-2)^5 - (3-5)}{3(2) + 4(-6)}$
13. $(5^2 - \sqrt{81})(-2^3 + 8)$
14. $36 \div 6 \cdot 2 \div 4 + 4$
15. $3^3 - 2(5 - 6 \cdot 3)$
16. $\dfrac{2}{3} + \dfrac{4}{5} - \dfrac{1}{6} - \dfrac{1}{3}$

Evaluate each expression if $x = -1$, $y = 2$ and $z = -3$.

17. $2x^2 + 8y - z$
18. $\dfrac{x+3}{4} - \dfrac{z^3}{y}$
19. $2z \div y \cdot x$
20. $3x^2 + 3xy + z^2$

Simplify each expression:

21. $11 - (x - 3)$
22. $\dfrac{2xh + h^2}{h}$
23. $\dfrac{2}{3}\left(\dfrac{15}{4}x - \dfrac{27}{14}y + \dfrac{9}{2}\right)$

Evaluate each expression.

24. $|5|$
25. $\left|-\dfrac{9}{3}\right|$
26. $-\left|\dfrac{3}{4}\right|$
27. $-\left|\dfrac{31}{3}\right|$

Let $x = 3$ and $y = -4$. Evaluate each expression.

28. $|x + y|$
29. $|x| + |y|$
30. $\dfrac{2|x| + 2|y|}{|xy|}$
31. $\dfrac{|3x - 2y|}{-|y|}$

Simplify each expression.

1. $2^3 2^2$

2. $(-3x^3)(2x^6)$

3. $(8y^2)(4y^{10})$

4. $(-2x^2y^3)^3$

5. $\left(\frac{x^5}{y^{10}}\right)^2$

Identify each expression as a polynomial or not a polynomial. For each polynomial, give the degree and identify it as a monomial, binomial, trinomial, or none of these.

6. $m^3 + 2m$

7. $x^2 + 4x + 4$

8. $6x^2y^3 + 11x^4y^2$

9. $\sqrt{2}m^3 - \sqrt{5}m^2 + 4m + 6$

10. $\frac{6}{x^2} + \frac{8}{x} + \frac{2}{3}$

Find each sum or difference.

11. $(3x^2 - 4x + 8) + (4x^2 + 5x - 1)$

12. $(2y^3 + 5y + 9) + (3y^2 - 2y + 5)$

13. $(4x^2 + 8x^5 - 7) - (4x^2 - 3x - 6)$

14. $(2x^2 + 4x - 1) - (3x^2 - 2x - 4) + (8x^2 + 11)$

Find each product.

15. $(2x + 3)(3x - 4)$

16. $(x - \frac{1}{2})(x + \frac{3}{4})$

17. $(-2y^2)(3y^3 - 6y^2 + 5y - 7)$

18. $(x + 3)(5x^4 + 2x^2 - 6x - 1)$

Find each product.

19. $(3x + 4)(3x - 4)$

20. $(2y + 3)^3$

21. $(3x - 6)^2$

22. $[(2x + 3) + 6][(2x + 3) - 6]$

Perform the indicated operations.

23. $2(2x^2 + 3x - 6) - 3(x^2 - 6x - 1)$

24. $x^3 \cdot 2x - (1 + x^2) \cdot 3x^2$

25. $-x^2(7x - 3) - 4x(2x + 3)$

26. $(4x^3 - 6x^2)(x - 1) - (x^4 - 2x^3)$

Factor out the greatest common factor from each polynomial.

1. $3x^4 - 18x^3 - 12x^2$

2. $10x^2 - 24x^4 + 12x^6$

3. $x^3 + 3x^2h + 3xh^2 + h^3 - x^3$

4. $(x-1)^3 + 4(x-1)^2 + 3(x-1)$

Factor each polynomial by grouping.

5. $xy - x + 4y - 4$

6. $6xy + 14x - 15y - 35$

7. $2y + 10 - 3xy - 15x$

8. $x^3 + 3x^2 + 4x + 12$

Factor each trinomial.

9. $2x^2y + 16xy + 30y$

10. $3x^2 + 5x + 2$

11. $5x^2 - 18x - 8$

12. $24x^4y + 30x^3y^2 - 9x^2y^3$

Factor each polynomial.

13. $9x^2 - 81y^2$

14. $25x^2 - 16$

15. $x^3 - 8$

16. $x^4 + 64x$

17. $(x+1)^3 - 64$

Factor each polynomial by substitution.

18. $3m^4 + 14m^2 + 8$

19. $10(x-1)^2 + (x-1) - 2$

20. $15(2-x)^2 + 2(2-x) - 24$

21. $3x^6 - 7x^3 - 40$

Find the domain of each rational expression.

1. $\dfrac{2x+1}{x-7}$

2. $\dfrac{x-2}{x^2+4x+3}$

3. $\dfrac{4}{x^2+2x+4}$

Write each rational expression in lowest terms.

4. $\dfrac{3x-9}{4x-12}$

5. $\dfrac{2x-4}{4x-8}$

6. $\dfrac{3x-6}{x^2-7x+10}$

7. $\dfrac{x^2-25x}{5x}$

8. $\dfrac{2x^2+x-3}{2x^2+7x+6}$

Find each product or quotient.

9. $\dfrac{8x^2}{24y} \div \dfrac{3x}{9y^2}$

10. $\dfrac{4x+6}{12} \cdot \dfrac{6}{2x+3}$

11. $\dfrac{x^2+xy}{20x-25} \cdot \dfrac{4x-5}{x^2}$

12. $\dfrac{x^2+2x-15}{x^2-9} \div \dfrac{2x^2+11x+5}{2x^2+7x+3}$

13. $\dfrac{x^2+x-2}{x^2+3x-4} \cdot \dfrac{x^2+2x-8}{x^2-3x-10}$

14. $\dfrac{4x^2-21x+5}{2x^2-x-1} \cdot \dfrac{3-2x-x^2}{x^2+2x+1} \div \dfrac{16x^2-1}{1-x^2}$

Perform each addition or subtraction.

15. $\dfrac{1}{4x} + \dfrac{2x}{7}$

16. $\dfrac{5x^2}{7} - \dfrac{7}{10x}$

17. $\dfrac{x^2+x}{x} + \dfrac{x+1}{2x-5}$

18. $\dfrac{x+1}{x} - \dfrac{x-5}{x+3}$

19. $\dfrac{x(3x-1)}{x^2+1} + \dfrac{x-1}{x+1}$

20. $\dfrac{1}{3x} + \dfrac{1}{5(2x-1)} - \dfrac{1}{15(x+2)}$

Simplify each expression.

21. $\dfrac{\frac{1}{x}+2}{4-\frac{1}{x}}$

22. $\dfrac{3+\frac{1}{x+4}}{\frac{1}{x-4}-1}$

Write each expression with only positive exponents and evaluate if possible. Assume all variables represent nonzero real numbers.

1. -2^{-3} 2. 7^{-3} 3. $3x^{-2}$ 4. $-6y^{-2}$

Perform the indicated operations. Write each answer using only positive exponents. Assume all variables represent nonzero real numbers.

5. $\frac{x^2}{x^8}$ 6. $\frac{y^{-3}}{y^{-4}}$ 7. $\frac{16x^{-2}y^{-3}}{4x^{-3}y^5}$ 8. $\frac{24x^3\left(y^3\right)^{-2}}{8x^4y^{-7}}$

Simplify each expression.

9. $81^{1/4}$ 10. $16^{1/2}$ 11. ${\frac{-8}{27}}^{1/3}$ 12. $(-81)^{1/2}$

Perform the indicated operations. Write each answer using only positive exponents. Assume all variables represent positive real numbers.

13. $16^{3/2}$ 14. $\frac{25^{-7/2}}{25^{-3}}$ 15. $-\frac{2}{3}x^{-2/3}\cdot x^{-1}$ 16. $\frac{\left(y^{2/3}\right)^4}{\left(y^{-1}\right)^{1/6}}$

17. ${\frac{x^4y^6}{16}}^{1/2}\ {\frac{8x^{-3}}{y^9}}^{1/3}$ 18. $\frac{x^{4/7}x^{1/14}x^{1/2}}{\left(x^9\right)^{-1/3}}$

Find each product. Assume that all variables represent positive real numbers.

19. $x^{3/2}\left(2x^{4/3}-6x^{1/7}\right)$ 20. $\frac{1}{3}x^{-2/3}\left(x+x^2\right)$

21. $\left(x^{1/2}-x^{1/3}\right)\left(x+x^{1/3}\right)$ 22. $\left(y^{1/3}+y^{-1/3}\right)^2$

Factor, using the given common factor. Assume all variables represent positive real numbers.

23. $x^{-16}-x^{-8},\quad x^{-16}$ 24. $y^{8/9}+27y^{-7/9},\quad y^{-7/9}$

25. $(x+1)^{-4/3}+(x+1)^{-1/3}+(x+1)^{1/3},\ (x+1)^{-4/3}$

26. $(2x-1)^{-7/5}+(2x-1)^{1/5}+(2x-1)^{3/5},\ (2x-1)^{-7/5}$

Write in radical form. Assume all variables represent positive real numbers.

1. $(6x-4y)^{2/3}$

2. $(5+2y)^{1/7}$

Write in exponential form. Assume all variables represent positive real numbers.

3. $\sqrt[7]{x^3}$

4. $-y\sqrt[4]{5x^9}$

Simplify each expression. Assume all variables represent positive real numbers.

5. $\sqrt{(2x-1)^2}$

6. $-\sqrt[4]{625x^{13}y^9}$

7. $-\sqrt[3]{\dfrac{81}{125}}$

8. $\sqrt{\dfrac{x^7y^{10}}{z^{11}}}$

9. $-\sqrt[3]{\dfrac{16x^5}{y^{13}z^8}}$

10. $\sqrt[4]{\sqrt{x}}$

Simplify each expression, assume all variables represent positive numbers.

11. $6\sqrt{18}-3\sqrt{2}+9\sqrt{8}$

12. $\sqrt[3]{3}+5\sqrt[3]{24}-9\sqrt[3]{81}$

13. $(\sqrt{6}+3)(\sqrt{6}-3)$

14. $(\sqrt{x}+\sqrt[3]{x})^2$

15. $\dfrac{\sqrt[4]{16x^5y^3}\,\sqrt[4]{x^3y^6}}{\sqrt[4]{81x^4y}}$

16. $\dfrac{9x\sqrt{x}-x}{\sqrt{x}}$

Rationalize the denominator of each radical expression. Assume all variables represent nonnegative numbers and that no denominators are 0.

17. $\dfrac{\sqrt{2}}{\sqrt{3}-\sqrt{2}}$

18. $\dfrac{1+\sqrt{6}}{\sqrt{6}+\sqrt{3}}$

19. $\dfrac{\sqrt{y}}{7-\sqrt{y}}$

20. $\dfrac{5x}{2-\sqrt{x+y}}$

R.1 Sets

1. {20, 21, 22, 23, 24, 25, 26, 27, 28, 29}
2. {4, 8, 12, 16, 20, 24, 28}
3. $\frac{1}{3}, \frac{1}{9}, \frac{1}{27}, \frac{1}{81}, \frac{1}{243}, \frac{1}{729}$
4. {51, 48, 45, 42, 39}
5. infinite
6. infinite
7. finite
8. infinite
9. $\in$
10. $\in$
11. $\notin$
12. $\notin$
13. true
14. false
15. true
16. true
17. false
18. true
19. true
20. false
21. $\varnothing$
22. U
23. C
24. $\{2,3,4,6,8,9,10\}$

R.2 Real Numbers and Their Properties

1. 5
2. $-3, -\frac{8}{2}, 5$
3. $-3, \frac{2}{3}, -\frac{8}{2}, \frac{1}{9}, 5$
4. $\pi, \sqrt{13}$
5. –8
6. –9
7. –48
8. 125
9. 7
10. 12
11. –13
12. $-\frac{17}{9}$
13. 0
14. 7
15. 53
16. $\frac{49}{30}$
17. 21
18. 14
19. 3
20. 6
21. $14 - x$
22. $2x + h$
23. $\frac{5}{2}x - \frac{9}{7}y + 3$
24. 5
25. 3
26. $-\frac{3}{4}$
27. $-\frac{31}{3}$
28. 1
29. 7
30. $\frac{7}{6}$
31. $\frac{17}{4}$

R.3 Polynomials

1. 32

2. $-6x^9$

3. $32y^{12}$

4. $-8x^6y^9$

5. $\dfrac{x^{10}}{y^{20}}$

6. polynomial, 3, binomial

7. polynomial, 2, trinomial

8. polynomial, 6, binomial

9. polynomial, 3, none of these

10. not a polynomial

11. $7x^2+x+7$

12. $2y^3+3y^2+3y+14$

13. $8x^5+3x-1$

14. $7x^2+6x+14$

15. $6x^2+x-12$

16. $x^2+\frac{1}{4}x-\frac{3}{8}$

17. $-6y^5+12y^4-10y^3+14y^2$

18. $5x^5+15x^4+2x^3-19x-3$

19. $9x^2-16$

20. $8y^3+36y^2+54y+27$

21. $9x^2-36x+36$

22. $4x^2+12x-27$

23. $x^2+24x-9$

24. $-x^4-3x^2$

25. $-7x^3-5x^2-12x$

26. $3x^4-8x^3+6x^2$

R.4 Factoring Polynomials

1. $3x^2(x^2-6x-4)$

2. $2x^2(5-12x^2+6x^4)$

3. $h(3x^2+3xh+h^2)$

4. $x(x-1)(x+2)$

5. $(x+4)(y-1)$

6. $(2x-5)(3y+7)$

7. $-(3x-2)(y+5)$

8. $(x+3)(x^2+4)$

9. $2y(x+3)(x+5)$

10. $(x+1)(3x+2)$

11. $(x-4)(5x+2)$

12. $3x^2y(2x+3y)(4x-y)$

13. $9(x-3y)(x+3y)$

14. $(5x-4)(5x+4)$

15. $(x-2)(x^2+2x+4)$

16. $x(x+4)(x^2-4x+16)$

17. $(x-3)(x^2+6x+21)$

18. $(m^2+4)(3m^2+2)$

19. $(2x-1)(5x-7)$

20. $(3x-10)(5x-4)$

21. $(x^3-5)(3x^3+8)$

R.5 Rational Expressions

1. $\{x \mid x \neq 7\}$

2. $\{x \mid x \neq -3, -1\}$

3. $(-\infty, \infty)$

4. $\frac{3}{4}$

5. $\frac{1}{2}$

6. $\frac{3}{x-5}$

7. $\frac{x-25}{5}$

8. $\frac{x-1}{x+2}$

9. xy

10. 1

11. $\frac{x+y}{5x}$

12. 1

13. $\frac{x-2}{x-5}$

14. $\frac{(x-5)(x+3)(x-1)}{(2x+1)(x+1)(4x+1)}$

15. $\frac{8x^2+7}{28x}$

16. $\frac{50x^3-49}{70x}$

17. $\frac{2x^2-2x-4}{2x-5}$

18. $\frac{9x+3}{x^2+3x}$

19. $\frac{4x^3+x^2-1}{x^3+x^2+x+1}$

20. $\frac{11x^2+22x-10}{30x^3+45x^2-30x}$

21. $\frac{1+2x}{4x-1}$

22. $\frac{3x^2+x-52}{-x^2+x+20}$

23. $\frac{x^2+x}{1-x}$

24. $\frac{x^3-x-1}{2x}$

R.6 Rational Exponents

1. $-\frac{1}{8}$

2. $\frac{1}{343}$

3. $\frac{3}{x^2}$

4. $\frac{-6}{y^2}$

5. $\frac{1}{x^6}$

6. y

7. $\frac{4x}{y^8}$

8. $\frac{3y}{x}$

9. 3

10. 4

11. $-\frac{2}{3}$

12. not a real number

13. 64

14. $\frac{1}{5}$

15. $-\frac{2}{3x^{5/3}}$

16. $y^{17/6}$

17. $\frac{x}{2}$

18. $x^{29/7}$

19. $2x^{17/6} - 6x^{23/14}$

20. $\frac{1}{3}x^{1/3} + \frac{1}{3}x^{4/3}$

21. $x^{3/2}+x^{5/6}-x^{4/3}-x^{2/3}$

22. $y^{2/3}+2+\dfrac{1}{y^{2/3}}$

23. $x^{-16}(1-x^{8})$

24. $y^{-7/9}(y^{5/3}+27)$

25. $(x+1)^{-4/3}\left(1+(x+1)+(x+1)^{5/3}\right)$

26. $(2x-1)^{-7/5}\left(1+(2x-1)^{8/5}+(2x-1)^{2}\right)$

R.7 Radical Expressions

1. $\sqrt[3]{(6x-4y)^{2}}$

2. $\sqrt[7]{5+2y}$

3. $x^{3/7}$

4. $-5^{1/4}x^{9/4}y$

5. $|2x-1|$

6. $-5x^{3}y^{2}\sqrt[4]{xy}$

7. $-\dfrac{3}{5}\sqrt[3]{3}$

8. $\dfrac{x^{3}y^{5}}{z^{5}}\sqrt{\dfrac{x}{z}}$

9. $-\dfrac{2x}{y^{4}z^{2}}\sqrt[3]{\dfrac{2x^{2}}{yz^{2}}}$

10. $x^{1/8}$

11. $33\sqrt{2}$

12. $-16\sqrt[3]{3}$

13. -3

14. $x+2\sqrt[6]{x^{5}}+\sqrt[3]{x^{2}}$

15. $\dfrac{2xy^{2}}{3}$

16. $9x-\sqrt{x}$

17. $\sqrt{6}+2$

18. $\dfrac{\sqrt{6}-\sqrt{3}+6-3\sqrt{2}}{3}$

19. $\dfrac{7\sqrt{y}+y}{49-y}$

20. $\dfrac{10x+5x\sqrt{x+y}}{4-x-y}$

Solve each equation.

1. $4x-9=11x-2$

2. $\frac{7}{8}x+6-2x=\frac{3}{4}$

3. $3[x-(6-3x)+5]=11x+8$

4. $0.5-0.8x=1.4x+0.6$

5. $\frac{4}{5}x-0.6x+0.3=x-9$

6. $5-(x+3)=3x+4(x-9)$

7. $\frac{1}{5}(x-3)=\frac{1}{2}(x-1)-\frac{2}{3}x$

8. $3(8+6x)=12$

9. $-0.2-(0.4x-0.3)=-1.8-0.6x$

Decide whether each equation is an identity, a conditional equation, or a contradiction. Give the solution set.

10. $6(4x-1)+3=2(4x-1+8x)-1$

11. $-0.4x+0.75+2(0.8+0.2x)=-0.6$

12. $3x-2+4(8x+9)=2(6+x)$

Solve each formula for the indicated variable. Assume that the denominator is not 0 if variables appear in the denominator.

13. $v^2=v_0^2+\frac{2Fd}{m}$ for F

14. $v=\pi r^2h$ for h

15. $A=P+Prt$ for r

16. $\beta=-\frac{1}{V}\cdot\frac{dV}{dP}$ for V

Solve each equation for *x*.

17. $5(3a+x)=ab-c$

18. $\frac{a}{x+4}=ab+c$

19. $a-4bx=(a+x)(a+3y)$

20. $a^2-ax+b^2=bx-c$

In the metric system of weights and measures, temperature is measured in degrees Celsius $(°C)$ instead of degrees Fahrenheit $(°F)$. To convert back and forth between the two systems, we use the equations $C=\frac{5}{9}(F-32)$ and $F=\frac{9}{5}C+32$. In each exercise, convert to the other system. Round answers to the nearest tenth of a degree if necessary.

21. $0°C$

22. $74°F$

Solve each problem.

1. The length of a calculus book is 2 inches more than the width. The perimeter is 36 inches. Find the width.

2. An avid gardener wants to make a circular flowerbed with a circumference of 32 feet. What should the radius be?

3. The perimeter of a rectangle is 82 feet. Find the length and width of the rectangle if the length is 7 feet more than the width.

4. The perimeter of an isosceles triangle is 28 inches. Each of the two equal sides measures 9 inches. What is the length of the third side?

5. Scott drives to work in the morning using the toll roads and he averages 60 mph. In the afternoon he drives home on free roads and he averages 45 mph. It took him 50 minutes longer to make the trip in the afternoon. How far does Scott live from work?

6. Rachel paddles her canoe upstream on the Econ River for 45 minutes. Coming downstream with the same boat speed took 30 minutes. If the current in this river is 3 mph, what was her boat speed?

7. Suppose that tuition costs $69.69 per credit hour and that student fees are fixed at $25.00. Antonio paid $1070.35 for his fall classes. How many credit hours did he sign up for?

8. How many gallons of a 9% peroxide solution must be mixed with 3 gallons of a 15% peroxide solution to obtain a 12% solution?

9. How much pure acid should be added to 16 ml of a 21% acid solution to increase the concentration to 30%?

10. Mr. Davis has $7000 to invest in two accounts. Part of the money will be invested at 3% and the rest will be invested at 4%. Find the amount invested at each rate if the total annual income from interest is $240.

11. Ms. Gordon wants to invest $9000 in stocks. She invests part of the money in a stock that pays 4% annual interest. The rest is invested in a business that pays 5% interest. Find the amount invested in each account if the total annual income from interest is $430.

12. Dan is 3 times as old as Mo, and in 6 years he will be twice as old. How old are Dan and Mo now?

13. A mother is 24 years older than her daughter. In 18 years, the mother's age will be twice that of her daughter. How old are they now?

Identify each number as real, complex, or pure imaginary.

1. 3

2. $4i$

3. $2-7i$

Write each number as the product of a real number and *i*.

4. $\sqrt{-16}$

5. $\sqrt{-7}$

6. $-\sqrt{-36}$

7. $-\sqrt{-120}$

Multiply or divide as indicated. Simplify each answer.

8. $\sqrt{-9}\sqrt{16}$

9. $\sqrt{-2}\sqrt{-32}$

10. $\dfrac{\sqrt{-48}}{\sqrt{12}}$

11. $\dfrac{-\sqrt{30}}{\sqrt{-90}}$

12. $\dfrac{\sqrt{-8}\sqrt{-4}}{\sqrt{2}}$

Write each number in standard form $a+bi$.

13. $\dfrac{-8+\sqrt{-48}}{4}$

14. $\dfrac{18-\sqrt{-72}}{6}$

Find each sum or difference. Write the answer in standard form.

15. $(4+5i)+(2+3i)$

16. $(7-6i)+(2-3i)$

17. $(2+5i)-(3+4i)$

18. $(4-3i)-(3-4i)$

Find each product. Write the answer in standard form.

19. $(5-2i)(3+4i)$

20. $(2+5i)^2$

21. $i(4-3i)^2$

22. $(1+i)(1-i)(2+3i)(2-3i)$

Simplify each power of *i*.

23. i^{28}

24. i^{29}

25. i^{-109}

Find each quotient. Write the answer in standard form.

26. $\dfrac{2+5i}{4+i}$

27. $\dfrac{-3+2i}{-1-4i}$

28. $\dfrac{-7}{-i}$

Solve each equation by the zero-factor property.

1. $x^2 + 2x = 8$

2. $3x^2 - 11x = -6$

Solve each equation by the square root property.

3. $x^2 = -36$

4. $(3x-1)^2 = -12$

Solve each equation by completing the square.

5. $x^2 + 2x + 4 = 0$

6. $x^2 + 14x + 24 = 0$

7. $18x^2 - 9x = 20$

8. $2x^2 - 6x + 1 = 0$

Solve each equation using the quadratic formula.

9. $x^2 - 6x - 27 = 0$

10. $5x^2 - 6x = 9$

11. $5x^2 + 6x = -9$

12. $x^2 + 7x - 3 = 0$

13. $x^2 + 2x - 4 = 0$

Solve each cubic equation.

14. $x^3 - 64 = 0$

15. $x^3 + 125 = 0$

16. $8x^3 = 27$

Solve each equation for the indicated variable. Assume no denominators are 0.

17. $F = \frac{km_1m_2}{d^2}$ for d

18. $A = P \ 1+\frac{r}{n}^{\ 2}$ for r

Evaluate the discriminant for each equation. Then use it to predict the number of distinct solutions, and whether they are rational, irrational or nonreal complex. Do not solve the equation.

19. $x^2 - 5x - 14 = 0$

20. $2x^2 - 7x - 8 = 0$

21. $3x^2 - 2x + 1 = 0$

For each pair of numbers, find the values of *a*, *b*, and *c* for which the quadratic equation $ax^2 + bx + c = 0$ has the given numbers as solutions.

22. $-6, 2$

23. $2+\sqrt{3},\ 2-\sqrt{3}$

24. $4i,\ -4i$

Solve each problem.

1. A farmer wants to set up a pigpen using 40 feet of fence to enclose a rectangular area of 51 square feet. Find the dimensions of the pigpen.

2. A vegetable gardener has a tilled piece of soil that measures 6 yards by 12 yards. She wants to put a border of marigolds of the same width within the four sides and leave an area of 40 square yards for vegetables. How wide can the border be?

3. A playground has an area of 456 meters and a perimeter of 100 meters. How long and wide is it?

4. A classroom has an area of 952 square feet. The length of the room is 6 feet more than its width. How long and wide is it?

5. Frank is decorating a porch that measures 10 yards by 20 yards. He wants to paint a uniform border around the sides and leave 96 square yards in the middle of the floor for a faux rug painting. How wide should he paint the border?

6. A rectangular piece of cardboard is 5 inches longer than it is wide. Three-inch squares are cut out of each corner and the resulting flaps are turned up and taped to form an open box. If the volume of the resulting box is 18 square inches, what were the original dimensions of the piece of cardboard?

7. A can of tuna has a surface area of 48.69 square inches. Its height is 1.87 inches. What is the radius of the can?

8. The volume of an 18-ounce box of cereal is 4524 cubic centimeters. The width of the box is 11.5 cm less than the length and the height is 29 cm. Find the length and width of the box.

9. The volume of a box of Jell-O is 9.7 cubic inches. The width of the box is 0.5 inches less than the length and its depth is 1 inch. Find the length and width of the box to the nearest thousandth.

10. A fish is being pulled onto a boat with a line attached to the fish at water level. When the fish is 10 feet from the boat, the length of the fishing line from the boat to the fish is 2 feet longer than the height of the fishing pole from the water. Find the height of the fishing pole.

11. A kite is flying on 20 feet of string. Its horizontal distance from the person flying it is 4 feet less than its vertical distance from the ground. Find its horizontal distance from the person and its vertical distance form the ground.

12. The size of a rectangular TV screen is given by the length of its diagonal. The length of the screen is 23 inches more than its width. Find the dimensions of a 63-inch TV screen.

Solve each equation.

1. $\dfrac{3}{x+1}=\dfrac{x}{x+1}+2$

2. $\dfrac{6}{x-3}-\dfrac{x+2}{x-3}=4$

3. $\dfrac{5}{x}+6=\dfrac{1}{x+2}$

4. $\dfrac{6}{x^2+x-2}+\dfrac{3}{x^2-x}=\dfrac{5}{x-1}$

5. $\dfrac{x}{x-4}-2=\dfrac{8}{x-4}$

6. $\dfrac{4}{x-2}-\dfrac{1}{2}=\dfrac{3x-8}{2x-4}$

7. $\dfrac{1}{x+1}=\dfrac{1}{3}-\dfrac{1}{6x+6}$

8. $\dfrac{4}{3x}+\dfrac{3}{3x+1}=-2$

9. $\dfrac{x+3}{x+4}+\dfrac{x+4}{x+3}=\dfrac{2x^2-16}{x^2+7x+12}$

10. $\dfrac{4x-8}{3x^{2/3}}=0$

Solve each equation.

11. $\sqrt{x-3}=x-5$

12. $\sqrt{x-2}-\sqrt{x}=7$

13. $\sqrt{4x^2+4}=2x+1$

14. $\sqrt{2x+9}-\sqrt{x+1}=\sqrt{x+5}$

15. $\sqrt{x+4}+\sqrt{2x-1}=3\sqrt{x-1}$

16. $5+\sqrt{x-6}=3$

17. $1+\sqrt[3]{x+3}=7$

18. $\sqrt[4]{3x+4}=\sqrt[4]{x-9}$

19. $\sqrt{17x-\sqrt{x^2-5}}=7$

20. $x^{5/3}(6-x)^{4/3}=0$

Solve each equation for the indicated variable. Assume all denominators are nonzero.

21. $A=2x\sqrt{r^2-x^2}$ for r

22. $\dfrac{1}{R}=\dfrac{1}{A}+\dfrac{1}{B}$ for B

23. $F=\dfrac{Gm_1m_2}{d^2}$ for d

24. $I=\dfrac{E}{r+R}$ for r

Solve each inequality. Write each solution set in interval notation.

1. $4x-1<x+8$
2. $3x+4<x+10$
3. $9(x+2)-4\le 10-x$
4. $\frac{1-2x}{3}+\frac{x}{2}\le x-\frac{1+3x}{6}$
5. $6(x+2)>2(2x+5)$

Solve each inequality. Write each solution set in interval notation.

6. $-11\le 3(x-1)-2\le 16$
7. $1<\frac{3-2x}{5}\le\frac{7}{3}$
8. $4<x+2\le 16$
9. $1<\frac{4x-3}{2}\le 8$
10. $-2\le\frac{10-x}{2}\le 6$

Solve each quadratic inequality. Write each solution set in interval notation.

11. $x^2-5x+6\le 0$
12. $x^2-x-12<0$
13. $x^2<100$
14. $x^2+4x>5$
15. $36x^2-24x\le 24$

Solve each rational inequality. Write each solution set in interval notation.

16. $3x-2<\frac{10-x}{2}$
17. $\frac{x+3}{2}\ge\frac{2-x}{-2}$
18. $\frac{4x-3}{2}\ge\frac{2-x}{3}$
19. $\frac{1}{x-6}\le 3$
20. $\frac{x+4}{x-2}\ge 2$

Solve each equation.

1. $|x-3|=6$

2. $|x-1|=9$

3. $\left|\dfrac{4x-7}{3}\right|=11$

4. $|9-4x|=|7+2x|$

5. $|3x-4|=|4-3x|$

Solve each inequality. Give the solution set using interval notation.

6. $|x-3|\le 1$

7. $|x-1|<3$

8. $2|x-6|>9$

9. $\left|\dfrac{1}{2}x-\dfrac{1}{3}\right|<4$

10. $|7x-9|\ge 15$

11. $|9x-2|-6\le 12$

12. $|2x+8|-9\ge 2$

13. $\left|\dfrac{15x-1}{5}\right|\le 3$

14. $|3x-4|-4\ge 9$

15. $|4x+3|-11\le 15$

Solve each equation or inequality.

16. $|7x+2|\ge 0$

17. $|9x-4|\ge -2$

18. $|4x+3|=0$

19. $|5x-2|\le 0$

20. $|6x+1|\le -3$

1.1 Linear Equations

1. {−1}

2. $\frac{14}{3}$

3. {11}

4. {−0.045}

5. {11.625}

6. $\frac{19}{4}$

7. $\frac{3}{11}$

8. $-\frac{2}{3}$

9. {−9.5}

10. identity

11. contradiction

12. conditional, $-\frac{2}{3}$

13. $F = \frac{m\left(v^2 - v_0^2\right)}{2d}$

14. $h = \frac{v}{\pi r^2}$

15. $r = \frac{A-P}{Pt}$

16. $V = -\frac{1}{\beta} \cdot \frac{dV}{dP}$

17. $x = \frac{ab-c-15a}{5}$

18. $x = \frac{a}{ab+c} - 4$

19. $x = \frac{a^2+3ay-a}{-4b-a-3y}$

20. $x = \frac{a^2+b^2+c}{a+b}$

21. 32°*F*

22. 23.3°*C*

1.2 Applications and Modeling with Linear Equations

1. 8 inches

2. 5.1 feet

3. 17 feet, 24 feet

4. 10 inches

5. 150 miles

6. 15mph

7. 15 credits

8. 3 gallons

9. 2.1 ml

10. \$4000 at 3%; \$3000 at 4%

11. \$2000 at 4%; \$7000 at 5%

12. Don is 18; Mo is 6

13. Mother is 30; daughter is 6

1.3 Complex Numbers

1. real, complex

2. pure imaginary

3. complex

4. $4i$

5. $\sqrt{7}i$

6. $-6i$

7. $-2\sqrt{30}i$

8. $12i$

9. −8

10. $2i$

11. $\frac{\sqrt{3}}{3}i$

12. −4

13. $-2+\sqrt{3}i$

14. $3-\sqrt{2}i$

15. $6+8i$

16. $9-9i$

17. $-1+i$

18. $1+i$

19. $23+14i$

20. $-21+20i$

21. $24+7i$

22. 26

23. 1

24. i

25. $-i$

26. $\frac{13}{17}+\frac{18}{17}i$

27. $-\frac{5}{17}-\frac{14}{17}i$

28. $-7i$

1.4 Quadratic Equations

1. $\{-4,2\}$

2. $\frac{2}{3},3$

3. $\{-6i,6i\}$

4. $\frac{1\pm 2\sqrt{3}i}{3}$

5. $\{-1\pm\sqrt{3}i\}$

6. $\{-12,-2\}$

7. $-\frac{5}{6},\frac{4}{3}$

8. $\frac{3\pm\sqrt{7}}{2}$

9. $\{-3,9\}$

10. $\frac{3\pm 3\sqrt{6}}{5}$

11. $-\frac{3}{5}\pm\frac{6}{5}i$

12. $\frac{-7\pm\sqrt{61}}{2}$

13. $\{-1\pm\sqrt{5}\}$

14. $\{4,-2\pm 2\sqrt{3}i\}$

15. $-5,\frac{5}{2}\pm\frac{5\sqrt{3}}{2}i$

16. $\frac{3}{2},-\frac{3}{4}\pm\frac{3\sqrt{3}}{4}i$

17. $d=\pm\sqrt{\frac{k\,m_1\,m_2}{F}}$

18. $r=-n\pm n\sqrt{\frac{A}{P}}$

19. 2 distinct rational solutions

20. 2 distinct irrational solutions

21. 2 distinct nonreal complex solutions

22. $a=1,b=4,c=-12$

23. $a=1,b=-4,c=1$

24. $a=1,b=0,c=16$

1.5 Applications and Modeling with Quadratic Equations

1. 17 feet by 3 feet

2. 1 yard

3. 12 m by 38 m

4. 28 feet by 34 feet

5. 2 yards

6. 7 in. by 12 in.

7. 2 inches

8. 19.5 cm by 8 cm

9. 2.875 in. by 3.375 in.

10. 24 feet

11. 12 feet by 16 feet

12. 31.5 in. by 54.5 in.

1.6 Other Types of Equations and Applications

1. $\frac{1}{3}$

2. $\frac{16}{5}$

3. $-\frac{5}{3},-1$

4. $-\frac{6}{5}$

5. $\{0\}$

6. $\frac{9}{2}$

7. $\frac{5}{2}$

8. $-\frac{1}{6},-\frac{4}{3}$

9. $-\frac{41}{14}$

10. $\{2\}$

11. $\{7\}$

12. $\varnothing$

13. $\frac{3}{4}$

14. $-\frac{1}{2}$

15. $\{5\}$

16. $\varnothing$

17. $\{213\}$

18. $-\frac{13}{2}$

19. $\{3\}$

20. $\{0,6\}$

21. $r=\pm\sqrt{\frac{A^2}{4x^2}+x^2}$

22. $B=\frac{AR}{A-R}$

23. $d=\pm\sqrt{\frac{G\,m_1\,m_2}{F}}$

24. $r=\frac{E-IR}{I}$

1.7 Inequalities

1. $(-\infty,3)$

2. $(-\infty,3)$

3. $-\infty,-\frac{2}{5}$

4. $\frac{3}{4},\infty$

5. $(-1,\infty)$

6. $[-2,7]$

7. $-\frac{13}{3},-1$

8. $(2,14]$

9. $\frac{5}{4},\frac{19}{4}$

10. $[-2,14]$

11. $[2,3]$

12. $(-3,4)$

13. $(-10,10)$

14. $(-\infty,-5)\cup(1,\infty)$

15. $\frac{1-\sqrt{7}}{3},\frac{1+\sqrt{7}}{3}$

16. $(-\infty,2)$

17. $(-\infty,\infty)$

18. $\frac{13}{14},\infty$

19. $(-\infty,6]\cup\ \frac{19}{3},\infty$

20. $[2,8]$

1.8 Absolute Value Equations and Inequalities

1. $\{-3,9\}$

2. $\{-8,10\}$

3. $-\frac{13}{2},10$

4. $\frac{1}{3},8$

5. $(-\infty,\infty)$

6. $[2,4]$

7. $(-2,4)$

8. $-\infty,\frac{3}{2} \cup \frac{21}{2},\infty$

9. $-\frac{22}{3},\frac{26}{3}$

10. $-\infty,-\frac{6}{7} \cup \frac{24}{7},\infty$

11. $-\frac{16}{9},\frac{20}{9}$

12. $-\infty,-\frac{19}{2} \cup \frac{3}{2},\infty$

13. $-\frac{14}{15},\frac{16}{15}$

14. $(-\infty,-3] \cup \frac{17}{3},\infty$

15. $-\frac{29}{4},\frac{23}{4}$

16. $(-\infty,\infty)$

17. $(-\infty,\infty)$

18. $-\frac{3}{4}$

19. $\frac{2}{5}$

20. $\varnothing$

For the points P and Q, find the distance $d(P,Q)$ and the coordinates of the midpoint of the segment PQ.

1. $(1,4),(-6,-1)$
2. $(-2,-5),(1,7)$
3. $(0,9),(2,19)$
4. $(-5,12),(1,-11)$

Determine whether the three points are the vertices of a right triangle.

5. $(-2,1),(2,3),(2,1)$
6. $(-3,3),(2,-2),(3,3)$
7. $(10,40),(30,-20),(10,-20)$
8. $(-75,-50),(75,-25),(25,25)$

Determine whether the three points are collinear.

9. $(-2,3),(0,3),(2,3)$
10. $(-2,0),(0,-2),(3,-2)$
11. $(1,2),(2,1),(-1,2)$
12. $(-3,-2),(0,0),(3,2)$

Find the coordinates of the other endpoint of each segment, given its midpoint and one endpoint.

13. midpoint $(-9,-3)$, endpoint $(-5,-7)$
14. midpoint $(-1,-1)$, endpoint $(2,-5)$
15. midpoint $(5,2)$, endpoint $(4,6)$
16. midpoint $(-4.5,4)$, endpoint $(-2,13)$

For each equation, give three ordered pairs that are solutions.

17. $x+y=7$
18. $4x=4y+16$
19. $y=x^2-6$
20. $y=-x^2$
21. $y=x^3+3$
22. $y=2-x^3$
23. $y=-|x-7|$
24. $y=\sqrt{x}+5$

Find the center-radius form of the equation of each circle.

1. center $(0,0)$; radius 5
2. center $(0,0)$; radius 1
3. center $(0,0)$; radius 3
4. center $(0,0)$; radius $\sqrt{7}$
5. center $(1,2)$; radius 1
6. center $(-3,-1)$; radius 2
7. center $(1,-3)$; radius 7
8. center $(-2,3)$; radius $\sqrt{5}$
9. center $(3,4)$; radius 3
10. center $(-5,-7)$; radius $\sqrt{2}$

Decide whether or not each equation has a circle as its graph. If it does, give the center and the radius.

11. $x^2 - 6x + y^2 + 2y - 15 = 0$
12. $x^2 - 2x + y^2 - 3 = 0$
13. $4x^2 + 9y^2 - 16x + 18y - 11 = 0$
14. $x^2 + 4x + y^2 - 6y = 3$
15. $x^2 + y^2 = 16$
16. $x^2 + y^2 - 12y = -32$
17. $x^2 + y^2 - 8x = -10$
18. $x^2 + y^2 + 2x - 6y = -6$
19. $x^2 + y^2 + 8x + 12y = -54$
20. $x^2 + y^2 - 4x - 6y = -13$

Decide whether each relation defines a function and give the domain and range.

1. $\{(-4,0),(-2,1),(0,2),(3,9)\}$

2. $\{(-2,7),(2,3),(-2,0),(5,-2)\}$

3. $\{(3.5,2.5),(2.0,2.0),(2.5,4.0),(3.5,3.5)\}$

4. $\{(50,10),(100,13),(150,14),(250,13)\}$

Decide whether each relation defines *y* as a function of *x*. Give the domain and range.

5. $y = x^6$

6. $y = \sqrt{4-2x}$

7. $2x - 2y < 8$

8. $x\,y = 2$

Let $f(x) = x^2 - 6x$ and $g(x) = 2x + 1$. Find the following.

9. $f(-3)$

10. $f(a+h)$

11. $g \; -\frac{1}{2}$

12. $g(3t-4)$

For each function, find $f(-1)$ and $f(3)$.

13. $f = \;\; -1,\frac{13}{3},(0,4),(3,3),(12,0)$

14. $f(x) = 3x^4 - 4x^3$

15. $f(x) = 1.75x$

16. $f(x) = 3 - |x|$

An equation that defines *y* as a function of *x* is given. Solve for *y* in terms of *x* and represent *y* with the function notation $f(x)$.

17. $3x - 4y = 5$

18. $y - 2x^2 = 4$

19. $x + 3y = 6$

20. $y + 6x^2 = 2$

Determine the intervals of the domain for which each function is (a) increasing, (b) decreasing, (c) constant.

21. $f(x) = |x| - 2$

22. $f(x) = \sqrt{x} - 3$

23. $f(x) = (x+1)^3$

24. $f(x) = (x-3)^2 + 4$

Graph each linear function. Identify any constant functions. Give the domain and the range.

1. $f(x) = x + 2$

2. $f(x) = -x - 1$

3. $f(x) = 4$

4. $f(x) = \frac{2}{3}x + \frac{5}{3}$

5. $f(x) = 3x$

Graph each vertical line. Give the domain and range of the relation.

6. $x = 2$

7. $x + 3 = 0$

8. $2x - 6 = 0$

Find the slope of the line satisfying the given conditions.

9. through $(-2,-8)$ and $(1,1)$

10. through $(-3,5)$ and $(1,5)$

11. through $\frac{2}{3}, -\frac{5}{2}$ and $\frac{1}{2}, -\frac{5}{3}$

12. horizontal, through $(3,-4)$

13. vertical, through $(-1,2)$

Find the slope of the line.

14. $y = -2x + 3$

15. $y = x - 1$

16. $2y - 4x = 8$

17. $4y + 3x = 12$

18. $-2y = 6x$

19. $y + 1 = 2(x - 6)$

20. $2(y - 8) = 6(x + 3)$

Write an equation for the line described. Give answers in standard form.

1. through $(1,3)$; $m=\frac{2}{3}$

2. through $(0,0)$; $m=0$

3. through $(7,-2)$ and $(-3,4)$

4. through $(6,1)$; undefined slope

Write an equation for the line described. Give answers in slope-intercept form (if possible).

5. through $(-3,-2)$; $m=6$

6. through $\frac{7}{8},\frac{3}{4}$ and $\frac{4}{5},\frac{1}{2}$

7. vertical; through $(2,7)$

8. horizontal; through $(-1,6)$

Give the slope and y-intercept of each line.

9. $x+2y=5$

10. $y=\frac{1}{2}x+4$

11. $4y-x=-12$

12. $3x-2y=4$

Write an equation in standard form and in slope-intercept form for the line described.

13. through $(1,2)$; parallel to $y-2x=3$

14. through $(2,-7)$; perpendicular to $y=-\frac{5}{3}x-2$

15. through $(3,-4)$; parallel to $y+5x=7$

16. through $(-10,3)$; perpendicular to $y=3$

Determine whether the three points are collinear by using slopes.

17. $(-2,5),(0,3),(2,1)$

18. $(2,4),(4,8),(8,16)$

19. $(-1,3),(-4,0),(-7,-3)$

20. $(6,2),(1,3),(5,7)$

Determine the intervals of the domain over which each function is continuous.

1. $f(x)=\begin{cases} x+2 & \text{for } x\le -1 \\ \frac{1}{2}x+2 & \text{for } x>-1 \end{cases}$

2. $f(x)=\begin{cases} \frac{5}{3}x+2 & \text{for } x\le 0 \\ 3-x & \text{for } x>0 \end{cases}$

3. $f(x)=\begin{cases} 3x-12 & \text{for } x<3 \\ \frac{1}{2}x+3 & \text{for } x\ge 3 \end{cases}$

4. $f(x)=\begin{cases} 3-\frac{1}{2}x & \text{for } x\le -2 \\ 2x+8 & \text{for } x>-2 \end{cases}$

For each piecewise defined function, find (a) $f(-4)$, (b) $f(-2)$, (c) $f(0)$, and (d) $f(2)$.

5. $f(x)=\begin{cases} x+2 & \text{for } x\le -1 \\ \frac{1}{2}x+2 & \text{for } x>-1 \end{cases}$

6. $f(x)=\begin{cases} \frac{5}{3}x+2 & \text{for } x\le -6 \\ 3-x & \text{for } x>-6 \end{cases}$

7. $f(x)=\begin{cases} \frac{3}{4}x & \text{for } x<-2 \\ -\frac{1}{2}x+5 & \text{for } -2\le x<1 \\ 6-3x & \text{for } x\ge 1 \end{cases}$

8. $f(x)=\begin{cases} 7-x & \text{for } x<-3 \\ 3 & \text{for } -3\le x\le 2 \\ -2x+2 & \text{for } x>2 \end{cases}$

9. $f(x)=\begin{cases} 3+x & \text{for } x<-2 \\ x & \text{for } -2\le x\le 0 \\ -2x & \text{for } x>0 \end{cases}$

10. $f(x)=\begin{cases} 4x+2 & \text{for } x<-3 \\ \frac{1}{4}x-2 & \text{for } -3\le x\le 1 \\ -x & \text{for } x>1 \end{cases}$

11. $f(x)=\begin{cases} 3x-12 & \text{for } x\le 0 \\ \frac{1}{2}x+3 & \text{for } x>0 \end{cases}$

12. $f(x)=\begin{cases} 3-\frac{1}{2}x & \text{for } x\le -2 \\ 2x+8 & \text{for } x>-2 \end{cases}$

13. $f(x)=\begin{cases} 5-3x & \text{for } x<1 \\ 4+3x & \text{for } x\ge 1 \end{cases}$

14. $f(x)=\begin{cases} -4x-2 & \text{for } x<-1 \\ x & \text{for } x\ge -1 \end{cases}$

15. $f(x)=\begin{cases} 3x-1 & \text{for } x<-2 \\ 3x+1 & \text{for } -2\le x\le 2 \\ -\frac{1}{3}x-1 & \text{for } x>2 \end{cases}$

16. $f(x)=\begin{cases} -4x & \text{for } x\le -1 \\ 3 & \text{for } -1<x<1 \\ 2x & \text{for } x\ge 1 \end{cases}$

Given the function *f(x)*, describe how to obtain the graph of *g(x)*.

1. $f(x)=x^2;\ g(x)=-3x^2$
2. $f(x)=x^2;\ g(x)=\frac{1}{3}x^2$
3. $f(x)=x^2;\ g(x)=(-3x)^2$
4. $f(x)=|x|;\ g(x)=2|x|$
5. $f(x)=|x|;\ g(x)=-2|x|$
6. $f(x)=\sqrt{x};\ g(x)=\frac{1}{2}\sqrt{x}$

Without graphing, determine whether each equation has a graph that is symmetric with respect to the x-axis, the y-axis, the origin, or none of these.

7. $y=2x^2+3$
8. $y=-3x^4$
9. $y^2=4-x^2$
10. $y=2x^3$
11. $y=2x+1$
12. $y=\sqrt{x^2-4}$

Decide whether each function is even, odd, or neither.

13. $y=x^2+2$
14. $y=3x^3-1$
15. $y=0.25x^4-x^2+1$
16. $y=x^5-x^4$
17. $y=x^2+|x|$
18. $y=|x|-x$

Given the function *f(x)*, describe how to obtain the graph of *g(x)*.

19. $f(x)=x^2;\ g(x)=x^2+1$
20. $f(x)=x^2;\ g(x)=x^2-3$
21. $f(x)=x^2;\ g(x)=(x+1)^2$
22. $f(x)=\sqrt{x};\ g(x)=\sqrt{x-3}$
23. $f(x)=|x|;\ g(x)=|x-1|+2$
24. $f(x)=x^3;\ g(x)=(x-2)^3$

Let $f(x) = 4x^2 + 2x$ and let $g(x) = 5x - 4$. Find each of the following.

1. $(f+g)(2)$
2. $(fg)(2)$
3. $\frac{f}{g}(1)$

For the pair of functions defined, find $f+g, f-g, fg$, and $\frac{f}{g}$. Give the domain of each.

4. $f(x) = 6x + 4$, $g(x) = x - 3$
5. $f(x) = x^2 - 1$, $g(x) = 2x^2 - x + 1$
6. $f(x) = \sqrt{2x+1}$, $g(x) = \frac{1}{x}$

Use the table to evaluate each expression, if possible, (a) $(f+g)(1)$, (b) $(fg)(-2)$, (c) $\frac{f}{g}(0)$.

7.

x	$f(x)$	$g(x)$
-2	4	2
-1	2	2
0	5	1
1	-4	6

8.

x	$f(x)$	$g(x)$
-2	4	1
-1	1	1
0	-1	-3
1	2	2

9.

x	$f(x)$	$g(x)$
-2	-1	0
-1	1	1
0	-2	0
1	-1	2

For each of the functions defined as follows, find (a) $f(x+h)$, (b) $f(x+h) - f(x)$, and (c) $\frac{f(x+h) - f(x)}{h}$.

10. $f(x) = 2x - 1$
11. $f(x) = x^3$
12. $f(x) = x^2 - 2$

Let $f(x) = 3x - 2$ and $g(x) = -2x - 1$. Find each function value.

13. $(f \circ g)(-1)$
14. $(g \circ f)(3)$
15. $(f \circ f)(4)$

Find $(f \circ g)(x)$ and $(g \circ f)(x)$ for each pair of functions.

16. $f(x) = 4x + 2$, $g(x) = 6 - 3x$
17. $f(x) = 2x + 2$, $g(x) = 3x^2 - x + 2$
18. $f(x) = 5x^2 + 4x$, $g(x) = \sqrt{x-1}$

2.1 Rectangular Coordinates and Graphs

1. $d=\sqrt{74};\ -\frac{5}{2},\frac{3}{2}$

2. $d=3\sqrt{17};\ -\frac{1}{2},1$

3. $d=2\sqrt{26};\ (1,14)$

4. $d=\sqrt{565};\ -2,\frac{1}{2}$

5. yes

6. no

7. yes

8. no

9. yes

10. no

11. no

12. yes

13. $(-13,1)$

14. $(-4,3)$

15. $(6,-2)$

16. $(-7,-5)$

17. Answers will vary.

18. Answers will vary.

19. Answers will vary.

20. Answers will vary.

21. Answers will vary.

22. Answers will vary.

23. Answers will vary.

24. Answers will vary.

2.2 Circles

1. $x^2+y^2=25$

2. $x^2+y^2=1$

3. $x^2+y^2=9$

4. $x^2+y^2=7$

5. $(x-1)^2+(y-2)^2=1$

6. $(x+3)^2+(y+1)^2=4$

7. $(x-1)^2+(y+3)^2=49$

8. $(x+2)^2+(y-3)^2=5$

9. $(x-3)^2+(y-4)^2=9$

10. $(x+5)^2+(y+7)^2=2$

11. circle, $r=5$; C = (3,-1)

12. circle, $r=2$; C = (1,0)

13. not a circle

14. circle, $r=4$; C = (-2,3)

15. circle, $r=4$; C = (0,0)

16. circle, $r=2$; C = (0,6)

17. circle, $r=\sqrt{6}$; C = (4,0)

18. circle, $r=2$; C = (-1,3)

19. not a circle

20. not a circle

2.3 Functions

1. Function, $D=\{-4,-2,0,3\};R=\{0,1,2,9\}$

2. Not a function

3. Not a function

4. Function, $D=\{50,100,150,250\}$; $R=\{10,13,14\}$

5. y is a function of x; $D=(-\infty,\infty);R=[0,\infty)$

6. y is a function of x; $D=(-\infty,2];R=[0,\infty)$

7. y is not a function of x; $D=(-\infty,\infty); R=(-\infty,\infty)$

8. y is a function of x; $D=R=(-\infty,0)\cup(0,\infty)$

9. 27

10. $a^2+2ah+h^2-6a-6h$

11. 0

12. $6t-7$

13. $f(-1)=\frac{13}{3}; f(3)=3$

14. $f(-1)=7; f(3)=135$

15. $f(-1)=-1.75; f(3)=5.25$

16. $f(-1)=2; f(3)=0$

17. $f(x)=\frac{3}{4}x-\frac{5}{4}$

18. $f(x)=2x^2+4$

19. $f(x)=-\frac{1}{3}x+2$

20. $f(x)=-6x^2+2$

21. **(a)** $[0,\infty)$ **(b)** $(-\infty,0]$ **(c)** none

22. **(a)** $[0,\infty)$ **(b)** none **(c)** none

23. **(a)** $(-\infty,\infty)$ **(b)** none **(c)** none

24. **(a)** $[3,\infty)$ **(b)** $(-\infty,3]$ **(c)** none

2.4 Linear Functions

1. $D=R=(-\infty,\infty)$

2. $D=R=(-\infty,\infty)$

3. $D=(-\infty,\infty); R=\{4\}$; constant function

4. $D=R=(-\infty,\infty)$

5. $D=R=(-\infty,\infty)$

6. $D=\{2\}; R=(-\infty,\infty)$

7. $D=\{-3\}; R=(-\infty,\infty)$

8. $D=\{3\}; R=(-\infty,\infty)$

9. $m=3$

10. $m=0$

11. $m=-5$

12. $m=0$

13. undefined

14. $m=-2$

15. $m=1$

16. $m=2$

17. $m=-\frac{3}{4}$

18. $m=-3$

19. $m=2$

20. $m=3$

2.5 Equations of Lines; Curve Fitting

1. $2x-3y=-7$

2. $y=0$

3. $3x+5y=11$

4. $x=6$

5. $y=6x+16$

6. $y=\frac{10}{3}x-\frac{13}{6}$

7. $x=2$

8. $y=6$

9. $m=-\frac{1}{2}; y-\text{int}=\frac{5}{2}$

10. $m=\frac{1}{2}; y-\text{int}=4$

11. $m=\frac{1}{4}; y-\text{int}=-3$

12. $m=\frac{3}{2}; y-\text{int}=-2$

13. $2x-y=0$; $y=2x$

14. $3x-5y=41$; $y=\frac{3}{5}x-\frac{41}{5}$

15. $5x+y=11$; $y=-5x+11$

16. $x=-10$

17. yes

18. yes

19. yes

20. no

2.6 Graphs of Basic Functions

1. $(-\infty,-1]\cup(-1,\infty)$

2. $(-\infty,0]\cup(0,\infty)$

3. $(-\infty,3)\cup[3,\infty)$

4. $(-\infty,\infty)$

5. (a) -2, (b) 0, (c) 2, (d) 3

6. (a) 7, (b) 5, (c) 3, (d) 1

7. (a) -3, (b) 6, (c) 5, (d) 0

8. (a) 11, (b) 3, (c) 3, (d) 3

9. (a) -1, (b) -2, (c) 0, (d) -4

10. (a) -14, (b) $-\frac{5}{2}$, (c) -2, (d) -2

11. (a) -24, (b) -18, (c) -12, (d) 4

12. (a) 5, (b) 4, (c) 8, (d) 12

13. (a) 17, (b) 11, (c) 5, (d) 10

14. (a) 14, (b) 6, (c) 0, (d) 2

15. (a) -13, (b) -5, (c) 1, (d) 7

16. (a) 16, (b) 8, (c) 3, (d) 4

2.7 Graphing Techniques

1. a vertical stretch by a factor of 3 and a reflection across the x-axis

2. a vertical shrink by a factor of $\frac{1}{3}$

3. a vertical stretch by a factor of 9

4. a vertical stretch by a factor of 2

5. a vertical stretch by a factor of 2 and a reflection across the x-axis

6. a vertical shrink by a factor of $\frac{1}{2}$

7. y-axis

8. y-axis

9. x-axis, y-axis, origin

10. origin

11. none of these

12. y-axis

13. even

14. neither

15. even

16. neither

17. even

18. neither

19. a translation of 1 unit upward

20. a a translation of 3 units downward

21. a a translation of 1 unit to the left

22. a a translation of 3 units to the right

23. a a translation of 2 units downward and 1 unit to the right

24. a a translation of 2 units to the right

2.8 Function Operations and Composition

1. 26

2. 120

3. 6

4.

$$(f+g)(x)=7x+1;\ D=(-\infty,\infty)$$
$$(f-g)(x)=5x+7;\ D=(-\infty,\infty)$$
$$(f\cdot g)(x)=6x^2-14x-12;\ D=(-\infty,\infty)$$
$$\frac{f}{g}(x)=\frac{6x+4}{x-3};\ D=(-\infty,3)\cup(3,\infty)$$

5.

$$(f+g)(x)=3x^2-x;\ D=(-\infty,\infty)$$
$$(f-g)(x)=-x^2+x-2;\ D=(-\infty,\infty)$$
$$(f\cdot g)(x)=2x^4-x^3-x^2+x-1;\ D=(-\infty,\infty)$$
$$\frac{f}{g}(x)=\frac{x^2-1}{2x^2-x+1};\ D=(-\infty,\infty)$$

6.

$$(f+g)(x)=\sqrt{2x+1}+\frac{1}{x};\ D= -\frac{1}{2},0\ \cup(0,\infty)$$

$$(f-g)(x)=\sqrt{2x+1}-\frac{1}{x};\ D= -\frac{1}{2},0\ \cup(0,\infty)$$

$$(f\cdot g)(x)=\frac{\sqrt{2x+1}}{x};\ D= -\frac{1}{2},0\ \cup(0,\infty)$$

$$\frac{f}{g}\ (x)=x\sqrt{2x+1};\ D= -\frac{1}{2},\infty$$

7. (a) 2, (b) 8, (c) 5

8. (a) 4, (b) 4, (c) $\frac{1}{3}$

9. (a) 1, (b) 0, (c) undefined

10. (a) $2x+2h-1$, (b) $2h$, (c) 2

11. (a) $x^3+3x^2h+3xh^2+h^3$, (b) $3x^2h+3xh^2+h^3$, (c) $3x^2+3xh+h^2$

12. (a) $x^2+2xh+h^2-2$, (b) $2xh+h^2$, (c) $2x+h$

13. 1

14. -15

15. 28

16. $(f\circ g)(x)=26-12x$; $(g\circ f)(x)=-12x$

17. $(f\circ g)(x)=6x^2-2x+6$; $(g\circ f)(x)=12x^2+22x+12$

18. $(f\circ g)(x)=5x-5+4\sqrt{x-1}$; $(g\circ f)(x)=\sqrt{5x^2+4x-1}$

Give the vertex, axis, domain, and range for each of the following parabolas.

1. $f(x)= \left(x-\frac{5}{2}\right)^2-\frac{1}{4}$
2. $f(x)=(x-3)^2$
3. $f(x)=x^2-6x-1$
4. $f(x)=x^2+2x+4$
5. $f(x)=2x^2-7x+3$
6. $f(x)=x^2-5x+10$
7. $f(x)=x^2+x-6$
8. $f(x)=x^2-2x-3$
9. $f(x)=-2x^2+5x+1$
10. $f(x)=4x^2-12x+9$
11. $f(x)=9x^2-16$
12. $f(x)=x^2-7x+12$

Modeling

13. A volleyball player hits a ball and it goes straight up into the air with an initial velocity of 16 feet per second from an initial height of 4 feet. Find the function that describes the height of the ball in terms of time t and determine the time at which the ball reaches its maximum height.

14. Find two numbers whose sum is 50 and whose product is a maximum.

15. A farmer has 36 feet of fence to build a pigpen. He is going to use one of the sides of his barn as a side to the rectangular enclosure. Determine a function A that represents the total area of the enclosed region. What is the maximum area that can be enclosed?

16. A square piece of poster board is going to be made into an open box by cutting 3-inch squares from each corner and folding up the sides. Determine a function V that represents the volume of the box.

17. A gardener wants to dig up a piece of the yard and make a rectangular vegetable garden with a border around the four sides and a piece of border splitting the region in half. She has 48 feet of border. Write a function A that represents the area of the garden. What is the maximum area that can be enclosed?

18. A toy rocket is launched into the sky with an initial velocity of 193 feet per second from a launching pad 6 feet in the air. Find the function that describes the height of the rocket in terms of time t and determine the time at which the ball reaches its maximum height.

19. A math book is thrown straight up from the ground with an initial velocity of 96 feet per second. Find the function that describes the height of the book in terms of time t and determine the time at which the book reaches its maximum height.

20. Suppose that x represents one of two positive numbers whose sum is 30. Determine a function f that represents the product of these two numbers. For what number(s) is the product a maximum?

Use synthetic division to perform each division.

1. $\dfrac{5x^3-3x+4}{x-2}$

2. $\dfrac{2x^3-3x^2+1}{x-3}$

3. $\dfrac{x^3+4x^2-7}{x+3}$

4. $\dfrac{3x^4-4x^2+4}{x+2}$

5. $\dfrac{2x^3+3x^2+4x+7}{x-\frac{1}{2}}$

6. $\dfrac{x^3+x}{x+1}$

Express *f(x)* in the form $f(x)=(x-k)\cdot q(x)+r$ for the given value of *k*.

7. $f(x)=x^2-x-6;\ k=3$

8. $f(x)=12x^2-20x+3;\ k=\dfrac{3}{2}$

9. $f(x)=x^2+11x+16;\ k=-8$

10. $f(x)=5x^4+5x^3+2x^2-x;\ k=-1$

11. $f(x)=2x^3-5x^2-6x+17;\ k=3$

12. $f(x)=3x^3-4x^2-6x+7;\ k=2$

For each polynomial function use the remainder theorem and synthetic division to find *f(k)*.

13. $k=-1;\ f(x)=x^3-x^2+4x+2$

14. $k=-\dfrac{1}{2};\ f(x)=-8x^5-7x^4+9x$

15. $k=1;\ f(x)=7x^4-4x^3+1$

16. $k=4;\ f(x)=x^3-2x^2-7x+2$

17. $k=3;\ f(x)=6x^3+8x^2-17x+6$

18. $k=-5;\ f(x)=6x^2-9x-6$

Use synthetic division to decide whether the given number *k* is a zero of the given polynomial.

19. $f(x)=x^3+3x^2-x-3;\ k=-1$

20. $f(x)=x^4-8x^3+22x^2-24x+9;\ k=2$

21. $f(x)=x^4+3x^3-x^2-3x;\ k=-3$

22. $f(x)=x^3+6x^2-x-30;\ k=-5$

23. $f(x)=x^5-4x^3;\ k=3$

24. $f(x)=x^4+x^3-8x^2-2x+12;\ k=2$

Use the factor theorem and synthetic division to decide whether the second polynomial is a factor of the first.

1. $5x^4 + 7x^3 - 6x^2 + 4x + 8;\ x + 2$

2. $x^3 - 5x^2 - 2x + 24;\ x - 3$

3. $8x^3 - 36x^2 + 54x - 27;\ x - 3$

4. $x^3 - 6x^2 + 11x - 6;\ x + 1$

Factor *f(x)* into linear factors given that *k* is a zero of *f(x)*.

5. $f(x) = x^3 - x^2 - 24x - 36;\ k = 6$

6. $f(x) = x^3 - 6x^2 - x + 30;\ k = 3$

7. $f(x) = x^3 - 9x^2 + 2x + 48;\ k = -2$

8. $f(x) = x^3 - 6x^2 + 3x + 10;\ k = -1$

For each polynomial function, one zero is given. Find all others.

9. $f(x) = 2x^3 + 3x^2 - 2x;\ -2$

10. $f(x) = x^3 + 4x^2 - 10x + 12;\ 1 + i$

11. $f(x) = x^3 - 5x^2 - 16x + 80;\ 5$

12. $f(x) = 2x^3 + 3x^2 + 2x + 3;\ i$

For each polynomial function, find all rational zeros and factor *f(x)*.

13. $f(x) = x^3 + 7x^2 - 28x + 20$

14. $f(x) = x^3 - 12x^2 - 55x + 150$

15. $f(x) = 5x^3 + 36x^2 - 33x - 8$

16. $f(x) = 4x^3 + 9x^2 - 81x + 54$

For each polynomial function, find all zeros and their multiplicities.

17. $f(x) = x^3 - 3x^2 - 9x - 5$

18. $f(x) = x^3 - 6x^2 + 32$

19. $f(x) = x\left(x - 3 + \sqrt{17}\right)(5x + 1)$

20. $f(x) = (5x - 8)^4(3x + 2)^2$

Find a polynomial function of degree 3 with real coefficients that satisfies the given conditions.

21. zeros of $6,\ -5,\ 8;\ f(1) = 420$

22. zeros of $1,\ 4,\ 10;\ f(2) = 48$

23. zeros of $-3,\ -4,\ 1;\ f(-1) = 12$

24. zeros of $10,\ -2,\ 1;\ f(-2) = 0$

Find a polynomial function of least degree having only real coefficients with zeros as given.

25. $1 - i$ and $2 + i$

26. $4 + i$ and $5 - i$

Describe the end behavior of the graph of each polynomial function.

1. $f(x) = 3x^3 - 2x^2 + x - 1$

2. $f(x) = 4x^6 - 5x^3 - 2x$

3. $f(x) = x^4 + 2x^3 - 6x^2 + 2$

4. $f(x) = 2x^5 + x^4 - 6x^3$

Use the intermediate value theorem for polynomials to show that each polynomial function has a real zero between the numbers given.

5. $f(x) = 3x^3 + 5x^2 - 5x + 1;\ -3$ and 2

6. $f(x) = 8x^3 - 6x^2 - 5x + 3;\ -1$ and 0

7. $f(x) = 4x^3 + 9x^2 - 10x + 3;\ -4$ and -3

8. $f(x) = 6x^4 + 4x^3 + 4x - 6;$ 0 and 1

Show that the real zeros of each polynomial function satisfy the given conditions.

9. $f(x) = 2x^3 + 10x^2 - 2x;$ no real zeros greater than 1

10. $f(x) = 3x^3 + 18x^2 - 24x;$ no real zeros greater than 2

11. $f(x) = 8x^3 + 2x - 30;$ no real zeros less than 1

12. $f(x) = 27x^3 + 3x + 68;$ no real zeros greater than -1

For the given polynomial function, approximate each zero as a decimal to the nearest tenth.

13. $f(x) = 2x^3 + 3x^2 + 4x - 6$

14. $f(x) = 4x^6 + 8x^5 + 2x^4 - 3x^3 - 5x^2 - 6x - 9$

15. $f(x) = 5x^3 + 9x^2 - 10x - 17$

16. $f(x) = 5x^4 - 6x^3 + 8x - 9$

Use a graphing calculator to find the coordinates of the turning points of the graph of each polynomial function in the given domain interval. Give answers to the nearest hundredth.

17. $f(x) = x^3 + 4x^2 - 23x - 35;\ [-1, 4]$

18. $f(x) = 2x^4 + 2x^3 - 3x^2 - 6;\ [0.1, 1]$

19. $f(x) = 5x^3 - 3x^2 + 7;\ [0.2, 1]$

20. $f(x) = x^4 - 3x^3 + 2x^2 - 1;\ [1, 2]$

Explain how the graph of each function can be obtained from the graph of $y = \frac{1}{x}$ or $y = \frac{1}{x^2}$.

1. $y = -\frac{4}{x^2}$

2. $y = \frac{1}{x-2}$

3. $y = \frac{1}{x} - 7$

4. $y = \frac{1}{x^2} + 2$

5. $y = \frac{1}{x-6}$

6. $y = \frac{1}{(x+2)^2}$

7. $y = \frac{1}{x-3} + 4$

8. $y = \frac{3}{(x+4)^2} - 2$

Give the equations of any vertical, horizontal, or oblique asymptotes for the graph of each rational function.

9. $y = \frac{1}{x-4}$

10. $y = \frac{5}{x^2 - x - 6}$

11. $y = \frac{2x}{2x+4}$

12. $y = \frac{x+3}{x-1}$

13. $y = \frac{x^2+9}{x^2-4}$

14. $y = \frac{x^2-9}{x-1}$

15. $y = \frac{3x^2}{2x^2+4}$

16. $y = \frac{x^2+11x+16}{x+8}$

Find a possible equation for the function with a graph having the given features.

17.

x - intercept : 1 $\quad$ y - intercept : $-\frac{1}{2}$

V.A.: $x = -2$ $\quad$ H.A.: $y = 1$

18.

x - intercept : 0 $\quad$ y - intercept : 0

V.A.: $x = 4$ $\quad$ H.A.: $y = 1$

Solve each variation problem.

1. If y varies directly as x, and $y = 30$ when $x = 8$, find y when $x = 4$.

2. Two yards of fabric cost $10.52. How much will 6.5 yards of fabric cost?

3. The distance a car can travel varies directly with the amount of gas in the tank. On the highway, a small SUV travels 212 miles on 13 gallons of gas. How many gallons of gas will be used on a 1036-mile trip?

4. The length, L, of a pendulum varies directly with the square of its period, p, the time required for the pendulum to make a complete swing back and forth. The pendulum on a clock is 6 inches long and has a period of 2 seconds. If a clock has a pendulum that is 8 inches long, how long is its period?

5. The distance a projectile will travel varies directly as the square of the initial velocity. A projectile with an initial velocity of 300 feet per second travels for 15,000 feet. What must the initial velocity be in order for the projectile to travel 60,000 feet?

6. If y varies inversely as x, and $y = 3$ when $x = 4$, find y when $x = 8$.

7. If y varies inversely as $\sqrt{x}$, and $y = \frac{1}{4}$ when $x = 4$, find y when $x = 169$.

8. The time in days, d, required to relandscape a park is inversely proportional to the number of people, n, landscaping. If 15 people can do the job in 4 days, how many days would it take 6 people to do the job?

9. The amount of current, C, that flows through a circuit varies inversely with the resistance, R, on the circuit. A rod with a resistance of 6 ohms draws 5 amps of current. Find the current when the resistance is 500 ohms.

10. The intensity of illumination, I, from a lamp varies inversely with the square of the distance, d, the object is from the lamp. If the intensity at 5 feet is 48 foot-candles, find the intensity at 6 feet.

11. If y is jointly proportional to x and the square root of z, and $y = -18.954$ when $x = -1.35$ and $z = 15.21$, find y when $x = 4$ and $z = 144$.

12. If y varies directly as x and inversely as z^2, and $y = 100$ when $x = 1$ and $z = 3$, find y when $x = 5$ and $z = 2$.

3.1 Quadratic Functions and Models

	Vertex	Axis	Domain	Range
1.	Vertex: $(2.5,-.25)$	Axis: $x=2.5$	Domain: $(-\infty,\infty)$	Range: $[-.25,\infty)$
2.	Vertex: $(3,0)$	Axis: $x=3$	Domain: $(-\infty,\infty)$	Range: $[0,\infty)$
3.	Vertex: $(3,-10)$	Axis: $x=3$	Domain: $(-\infty,\infty)$	Range: $[-10,\infty)$
4.	Vertex: $(-1,3)$	Axis: $x=-1$	Domain: $(-\infty,\infty)$	Range: $[3,\infty)$
5.	Vertex: $(1.75,-3.125)$	Axis: $x=1.75$	Domain: $(-\infty,\infty)$	Range: $[-3.125,\infty)$
6.	Vertex: $(2.5,3.75)$	Axis: $x=2.5$	Domain: $(-\infty,\infty)$	Range: $[3.75,\infty)$
7.	Vertex: $(-.5,-6.25)$	Axis: $x=-.5$	Domain: $(-\infty,\infty)$	Range: $[-6.25,\infty)$
8.	Vertex: $(1,-4)$	Axis: $x=1$	Domain: $(-\infty,\infty)$	Range: $[-4,\infty)$
9.	Vertex: $(1.25,4.125)$	Axis: $x=1.25$	Domain: $(-\infty,\infty)$	Range: $(-\infty,4.125]$
10.	Vertex: $(1.5,0)$	Axis: $x=1.5$	Domain: $(-\infty,\infty)$	Range: $[0,\infty)$
11.	Vertex: $(0,-16)$	Axis: $x=0$	Domain: $(-\infty,\infty)$	Range: $[-16,\infty)$
12.	Vertex: $(3.5,-.25)$	Axis: $x=3.5$	Domain: $(-\infty,\infty)$	Range: $[-.25,\infty)$

13. $h(t)=-16t^2+16t+4$**;** maximum height after 0.5 second

14. 25 and 25

15. $A(x)=x(36-2x)$; maximum area of 162 square feet

16. $V(x)=3(x-6)^2$

17. $A(x)=x(24-\frac{3}{2}x)$; maximum area of 96 square feet

18. $h(t)=-16t^2+193t+6$; maximum height after 6.03 seconds

19. $h(t)=-16t^2+96t$; maximum height after 3 seconds

20. $f(x)=x(30-x)$; $x=15$

3.2 Synthetic Division

1. $5x^2+10x+17+\frac{38}{x-2}$

2. $2x^2+3x+9+\frac{28}{x-3}$

3. $x^2+x-3+\frac{2}{x+3}$

4. $3x^3-6x^2+8x-16+\frac{36}{x+2}$

5. $2x^2+4x+6+\frac{10}{x-\frac{1}{2}}$

6. $x^2-x+2-\frac{2}{x+1}$

7. $f(x)=(x-3)(x+2)$

8. $f(x)=(x-\frac{3}{2})(12x-2)$

9. $f(x)=(x+8)(x+3)-8$

10. $f(x)=(x+1)(5x^3+2x-3)+3$

11. $f(x)=(x-3)(2x^2+x-3)+8$

12. $f(x)=(x-2)(3x^2+2x-2)+3$

13. –4

14. $-\frac{75}{16}$

15. 4

16. 6

17. 189

18. 189

19. yes

20. no

21. yes

22. yes

23. no

24. yes

3.3 Zeros of Polynomial Functions

1. yes

2. yes

3. no

4. no

5. $(x-6)(x+2)(x+3)$

6. $(x-5)(x-3)(x+2)$

7. $(x-8)(x-3)(x+2)$

8. $(x-5)(x-2)(x+1)$

9. $x=\frac{1}{2},0,-2$

10. $x=1+i,1-i,-6$

11. $x=5,4,-4$

12. $x=i,-i,-3/2$

13. $(x-2)(x-1)(x+10)$

14. $(x-15)(x-2)(x+5)$

15. $(x-1)(x+8)(5x+1)$

16. $(x-3)(x+6)(4x-3)$

17. $x=5$ multiplicity 1; $x=-1$ multiplicity 2

18. $x=-2$ multiplicity 1; $x=4$ multiplicity 2

19. $x=0,3-\sqrt{17},-1/5$

20. $x=\frac{8}{5}$ multiplicity 4; $x=-\frac{2}{3}$ multiplicity 2

21. $f(x)=2x^3-18x^2-44x+480$

22. $f(x)=3x^3-45x^2+162x-120$

23. $f(x)=-x^3-6x^2-5x+12$

24. $f(x)=x^3-9x^2-12x+20$

25. $f(x)=x^4-6x^3+15x^2-18x+10$

26. $f(x)=x^4-18x^3+123x^2-378x+442$

3.4 Polynomial Functions: Graphs, Applications, and Models

1. as $x\to-\infty, f(x)\to-\infty$, and as $x\to\infty, f(x)\to\infty$

2. as $x\to-\infty, f(x)\to\infty$, and as $x\to\infty, f(x)\to\infty$

3. as $x\to-\infty, f(x)\to\infty$, and as $x\to\infty, f(x)\to\infty$

4. as $x\to-\infty, f(x)\to-\infty$, and as $x\to\infty, f(x)\to\infty$

5. $f(-3)=-20<0;\ f(2)=35>0$

6. $f(-1)=-6<0;\ f(0)=3>0$

7. $f(-4)=-69<0;\ f(-3)=6>0$

8. $f(0)=-6<0;\ f(1)=8>0$

9. zeros = .19, 0, −5.19

10. zeros = 1.12, 0, −7.12

11. zero = 1.5

12. zero = −1.33

13. zero = .8

14. zeros = 1.1, −1.7

15. zeros = 1.4, −1.3, −1.9

16. zeros = 1.2, −1.2

17. (1.74,−57.64)

18. (.57,−6.39)

19. (.4,6.84)

20. (1.64,−1.62)

3.5 Rational Functions: Graphs, Applications, and Models

1. a vertical stretch by a factor of 4 and a reflection across the x-axis

2. a shift of 2 units to the right

3. a shift of 7 units down

4. a shift of 2 units up

5. a shift of 6 units to the right

6. a shift of 2 units to the left

7. a shift of 3 units to the right and a shift of 4 units up

8. a vertical stretch by a factor of 3, a shift of 4 units to the left and a shift of 2 units down

9. V.A. : $x = 4$; H.A. : $y = 0$

10. V.A. : $x = 3$ and $x = -2$; H.A. : $y = 0$

11. V.A. : $x = -2$; H.A. : $y = 1$

12. V.A. : $x = 1$; H.A. : $y = 1$

13. V.A. : $x = 2$ and $x = -2$; H.A. : $y = 1$

14. V.A. : $x = 1$; O.A. : $y = x + 1$

15. V.A. : none; H.A. : $y = 1.5$

16. V.A. : $x = -8$; O.A. : $y = x + 3$

17. $y = \frac{x-1}{x+2}$

18. $y = \frac{x}{x-4}$

3.6 Variation

1. $y = 15$

2. \$34.19

3. 63.5 gallons

4. 2.3 seconds

5. 600 feet per second

6. $y = \frac{3}{2}$

7. $y = \frac{1}{26}$

8. 10 days

9. 0.06 amps

10. 33.3 foot-candles

11. 172.8

12. 1125

Decide whether each function as defined is one-to-one.

1. $f(x)=3x+6$
2. $f(x)=\dfrac{1}{x^2-1}$
3. $f(x)=x^2-5$
4. $f(x)=\sqrt[3]{x-1}$
5. $f(x)=\sqrt{x-5}$

Determine whether each pair of functions defined as follows are inverses of each other.

6. $f(x)=4x+10;\ g(x)=\dfrac{1}{4}x-\dfrac{5}{2}$
7. $f(x)=x^2-2;\ g(x)=\sqrt{x+2}$
8. $f(x)=x^3-4;\ g(x)=\sqrt[3]{x+4}$
9. $f(x)=\dfrac{1}{x^2-3};\ g(x)=-\sqrt{\dfrac{1}{x}+3}$
10. $f(x)=\dfrac{3}{x-1};\ g(x)=\dfrac{3}{x}+1$

For each function defined as follows that is one-to-one, write an equation for the inverse function.

11. $f(x)=3x+9$
12. $f(x)=3x^2-6x$
13. $f(x)=x^3-1$
14. $f(x)=\dfrac{1}{x+6}$
15. $f(x)=\dfrac{1}{x^2}-2$
16. $f(x)=\sqrt{x-2}$
17. $f(x)=(6x-1)^2$
18. $f(x)=\sqrt[3]{x}+5$
19. $f(x)=(x+1)^3$
20. $f(x)=\sqrt{5-x}$

If $f(x) = 2^x$ and $g(x) = \frac{1}{3}^x$, find each of the following.

1. $f(1)$

2. $f(-2)$

3. $f(6.47)$

4. $g(0)$

5. $g(-1)$

6. $g(0.61)$

Solve each equation.

7. $2^{x+2} = 8^{4/3}$

8. $9 \cdot 3^{x-3} = 27^{-2x}$

9. $2^{x^2-x-4} = 4$

10. $25^{2-3x} = 125^{x-5}$

11. $\frac{1}{10}^x = 1000^{2x+4}$

12. $e^x = e^{6x-9}$

13. $\sqrt{e} = e^{2x}$

14. $x^{2/3} = 11$

Solve each problem involving compound interest.

16. Find the future value and interest earned if \$13,562 is invested at 2% compounded quarterly for 10 years.

17. Find the present value of \$8000 if interest is 4% compounded semi-annually for 2 years.

18. Find the required annual interest rate for \$5000 to grow to \$5261, if interest is compounded monthly for 12 months.

19. How long will it take for \$3000 to grow to \$4000 at an interest rate of 6% compounded continuously?

Use a graphing calculator to find the solution set of each equation.

20. $3^{4x-1} = x + 2$

21. $e^{2x+1} = x$

22. $x = 3^x - 3$

For each statement, write an equivalent statement in logarithmic or exponential form.

1. $\log_{10} 0.01 = -2$

2. $9^{1/2} = 3$

Solve each logarithmic equation.

3. $x = \log_2 \frac{1}{4}$

4. $\log_{10} 1 = x$

5. $x = 3^{\log_3 27}$

6. $\log_x 64 = 6$

7. $\log_5 x = 2$

8. $x = \log_2 \sqrt[3]{32}$

Write each expression as a sum, difference, or product of logarithms. Simplify the result if possible. Assume all variables represent positive real numbers.

9. $\log_2 \frac{x}{yz}$

10. $\log_b \sqrt{\frac{x^2}{y^3}}$

11. $\log_m \sqrt[3]{\frac{x^2 y^5}{z^7}}$

12. $\log_7 (x^2 - 2y^2)$

Write each expression as a single logarithm with coefficient 1. Assume that all variables represent positive real numbers.

13. $2\log_2 x + 2\log_2 y - \frac{1}{2}\log_2 z$

14. $\frac{1}{3}\log_m x - \frac{1}{4}\log_m y$

15. $\log_2 (x^2 - 9) - \log_2 (x - 3)$

Given $\log_{10} 3 = 0.4771$ and $\log_{10} 7 = 0.8451$, find each logarithm without using a calculator.

16. $\log_{10} 21$

17. $\log_{10} \frac{7}{3}$

18. $\log_{10} 63$

19. $\log_{10} \frac{27}{49}$

Use a calculator with logarithm keys to find an approximation for each expression. Give answers to four decimal places.

1. $\log(14)$
2. $\log(0.196)$
3. $\ln(30.6)$
4. $\ln\left(3 \cdot e^2\right)$

For each substance, find the pH from the given hydronium ion concentration.

5. ammonia, 3.16×10^{-12}
6. battery acid, 3.16×10^{-1}

Find the $[H_3O^+]$ for each substance with the given pH.

7. cola, 2.5
8. bleach, 12.5

Use the change-of-base theorem to find an approximation for each logarithm. Give answers to 4 decimal places.

9. $\log_3 24$
10. $\log_2 3$
11. $\log_8 2.7$
12. $\log_{\sqrt{2}} 9$

Given $f(x) = e^x$; $g(x) = 2^x$; $h(x) = \ln x$; $j(x) = \log_5 x$, evaluate the following.

13. $f(\ln 6)$
14. $g(\log_2 \sqrt{3})$
15. $h(e^{3\ln 5})$
16. $j(5^{2\log_5 5})$

Use the properties of logarithms to describe how the graph of the given function compares to the graph of $f(x) = \ln x$.

17. $f(x) = \ln\left(e^{1/3}x\right)$
18. $f(x) = \ln\left(x^{-1}\right)$
19. $f(x) = \ln(x) - \ln(e^4)$
20. $f(x) = \ln(x^3)$

Solve each equation. Round answers to 4 decimal places.

1. $10^{x+7} = 13$
2. $2^{2x-1} = 9$
3. $25^{x} = 10^{-3x}$
4. $e^{2x} = 4e^{x}$
5. $4^{2x+1} = 6^{x}$
6. $(x-2)e^{-x} = 0$
7. $\log_2(x+1) + \log_2(x-3) = 1$
8. $\log_3(x+2) - \log_3(x) = 3$
9. $\log x + \log(x-4) = 6$
10. $3\log x - 5 = 4$
11. $2\ln(4x) = 16$
12. $\log(x^2 - 9) - \log(x+3) = 1$

Solve each equation for the indicated variable. Use logarithms to the appropriate bases.

13. $N = 225e^{0.02t}$ for t
14. $y = -301\ln\dfrac{x}{207}$ for x
15. $R = 10 \cdot 2^{-t/5700}$ for t
16. $y = Ce^{kt}$ for t

Use a graphing calculator to solve each equation. Give irrational solutions correct to the nearest hundredth.

17. $x - 4 = \ln x$
18. $\ln(3 - e^{x}) = 5$
19. $\log x = 0.5x - 2$
20. $4 = x + \log_4 x$

Solve each problem.

1. The population of a town is $P = 95{,}700e^{0.013t}$ where $t = 0$ represents 1999. When will the population reach 200,000?

2. The number of flies in a lab is $N = 75e^{0.03t}$ where t is the number of days. When will the number of flies double?

3. The number of ants in a colony is $N = 100e^{0.012t}$ where t is the number of days. When will there be 600 ants?

4. A plasma TV depreciates according to $P = 3600e^{-0.19t}$ where t is the age of the plasma TV in years. How much will the TV be worth in 5 years?

5. The magnitude, M, of an earthquake is measured with the Richter scale, $M = \log_{10} I$ where I is the intensity of the earthquake. Find the intensity of the 1971 earthquake in Los Angeles that had a magnitude of 6.7.

6. The level of sound is measured in decibels and is modeled by the formula $D = 10\log_{10} \dfrac{I}{10^{-12}}$ where I is the intensity of the sound in watts per square meter. Find the decibel level for a noise with an intensity of $10^{-2.5}$ watt per square meter.

7. Refer back to exercise 6. What is the intensity of a sound with a decibel level of 89?

8. Find the doubling time of an investment earning 6% interest if interest is compounded continuously.

9. How long will it take \$30,000 to grow to \$75,000 if it is invested at 5.5% interest compounded continuously?

10. Find the amount of money you would need to invest now at 8% compounded continuously to have \$100,000 in 10 years.

11. Find the tripling time of an investment earning 9% interest compounded continuously.

12. What interest rate would give you \$90,000 if you invest \$50,000 for 20 years in an account that compounds interest continuously?

4.1 Inverse Functions

1. yes

2. no

3. no

4. yes

5. yes

6. yes

7. no

8. yes

9. no

10. yes

11. $f^{-1}(x)=\frac{1}{3}x-3$

12. not one-to-one

13. $f^{-1}(x)=\sqrt[3]{x+1}$

14. $f^{-1}(x)=\frac{1}{x}-6$

15. not one-to-one

16. $f^{-1}(x)=x^2+2; x\geq 0$

17. not one-to-one

18. $f^{-1}(x)=(x-5)^3$

19. $f^{-1}(x)=\sqrt[3]{x}-1$

20. $f^{-1}(x)=5-x^2; x\geq 0$

4.2 Exponential Functions

1. 2

2. $\frac{1}{4}$

3. 88.647

4. 1

5. 3

6. 0.51163

7. 2

8. $\frac{1}{7}$

9. –2, 3

10. $\frac{19}{9}$

11. $-\frac{12}{7}$

12. $\frac{9}{5}$

13. $\frac{1}{4}$

14. $-11^{3/2}, 11^{3/2}$

15. \$16,556.40, \$2994.41

16. \$7390.76

17. 5.1%

18. 4.8 years

19. 0.45, -2

20. no solution

21. –2.96, 1.34

22. 0.56

4.3 Logarithmic Functions

1. $10^{-2} = 0.01$

2. $\log_9 3 = \frac{1}{2}$

3. –2

4. 0

5. 27

6. 2

7. 25

8. $\frac{5}{3}$

9. $\log_2 x - \log_2 y - \log_2 z$

10. $\log_b x - \frac{3}{2}\log_b y$

11. $\frac{2}{3}\log_m x + \frac{5}{3}\log_m y - \frac{7}{3}\log_m z$

12. $\log_7(x^2 - 2y^2)$

13. $\log_2 \frac{x^2y^2}{\sqrt{z}}$

14. $\log_m \frac{\sqrt[3]{x}}{\sqrt[4]{y}}$

15. $\log_2(x+3)$

16. 1.3222

17. 0.3680

18. 1.7993

19. –0.2589

4.4 Evaluating Logarithms and the Change-of-Base Theorem

1. 1.1461

2. –0.7077

3. 3.4210

4. 3.0986

5. 11.5

6. 0.5

7. 3.16×10^{-3}

8. 3.16×10^{-13}

9. 2.8928

10. 1.5850

11. 0.4777

12. 6.3399

13. 6

14. $\sqrt{3}$

15. $3\ln 5$

16. 2

17. shifted up $\frac{1}{3}$ unit

18. reflection about the x-axis

4.5 Exponential and Logarithmic Equations

1. −5.8861

2. 2.0850

3. 0

4. 1.3863

5. −1.4134

6. 2

7. 3.4495

8. 0.0769

9. 1002.0020

10. 1000

11. 745.2395

12. 13

13. $t = 50 \ln \dfrac{N}{225}$

14. $x = 207e^{-y/301}$

15. $t = -5700 \log_2 \dfrac{R}{10}$

16. $t = \dfrac{1}{k} \ln \dfrac{y}{C}$

17. 5.75, 0.02

18. no solution

19. 5.48, 0.01

20. 3.17

4.6 Applications and Models of Exponential Growth and Decay

1. 2055

2. 23 days

3. 149 days

4. $1392.27

5. 5.01×10^6

6. 95 decibels

7. 7.9×10^{-4} w/m^2

8. 11.55years

9. 16.7 years

10. $44,932.90

11. 12.2 years

12. 2.9%

Solve each system by substitution.

1. $x + y = 16$
   $y = 7x$

2. $y = 2x - 8$
   $4y + 12 = 2x$

3. $x + y = 4$
   $2x + 3y = 9$

4. $3x + y = 9$
   $2x - 2y = 12$

Solve each system by elimination.

5. $2x + 4y = -12$
   $3x + 2y = 1$

6. $3x + y = 7$
   $2x - 5y = -1$

7. $x - y = 6$
   $2x - 3y = 9$

8. $2x + 3y = 11$
   $3x - y = 4$

Solve each system. State whether it is inconsistent or has infinitely many solutions.

9. $2x - y = 1$
   $8x - 4y = 3$

10. $2x - 5y = 6$
    $4x = 12 + 10y$

11. $6x + 2y = 1$
    $12x = 2 - 4y$

12. $2x + 3y = 2$
    $4x + 6y = 7$

Use a graphing calculator to solve each system. Express solutions to the nearest thousandth.

13. $5.8x - 2.5y = 6.44$
    $3.7x + 2.3y = 10.29$

14. $\sqrt{2}x - 4y = 9$
    $10x + 11y = 12$

15. $0.3x - \sqrt{2}y = 2$
    $\sqrt{7}x + 0.5y = 1$

16. $0.9x + 0.3y = 1.03$
    $2.5x - 8.5y = 11.1$

Solve each system of equations in three variables.

17. $x + 2y + 3z = 9$
    $2x - y + 2z = 11$
    $3x + 4y - 2z = -4$

18. $2x - 2y + 3z = 1$
    $2x - 6y - 4z = -9$
    $x + y + z = 6$

Write the augmented matrix for each system and give its size. Do not solve the system.

1. $5x-2y=-10$
   $4x+3y=-1$

2. $3x+6y-3z=-1$
   $-4x-3y-4z=3$
   $2x+2y+4z=2$

Write the system of equations associated with each augmented matrix. Do not solve the system.

3. $\left[\begin{array}{ccc|c} 1 & 1 & -1 & 2 \\ 2 & -1 & 1 & 6 \\ -1 & 2 & -1 & 1 \end{array}\right]$

4. $\left[\begin{array}{ccc|c} 3 & -4 & 8 & 0 \\ 5 & 3 & -4 & -2 \\ -5 & 3 & -8 & -3 \end{array}\right]$

Use the Gauss-Jordan method to solve each system of equations.

5. $-x+4y=9$
   $3x-2y=9$

6. $4x-6y=2$
   $2x-2y=-1$

7. $x+2y=9$
   $3x+4y=-4$

8. $2x\text{-}3y\text{=}6$
   $5x+y=1$

9. $x+5y+7z=2$
   $3x+9y-2z=3$
   $6y-4z=1$

10. $2x+3y-z=-8$
    $x+2y+2z=3$
    $-3x+y+z=-5$

11. $3x-y-z=-2$
    $2x+2y+3z=3$
    $-3x+2y+2z=5$

12. $-x+y-4z=12$
    $-8x-20y+4z=15$
    $2x+5y-z=12$

13. $2x-y+3z=-2$
    $y+2z=1$
    $3x-2y+z=3$

14. $-x+2y-3z=1$
    $2x+y+z=0$
    $x+3y+2z=4$

15. $-x-2y-3z=3$
    $2x+4y+z=2$
    $-x+2y=1$

16. $2x-y+3z=2$
    $x+2y+3z=-1$
    $3x-2y+z=3$

17. $x+y+z=4$
    $x-z=0$
    $y-z=7$

18. $x+z=3$
    $2x+3y+z=2$
    $x+3y+2z=3$

Find the value of each determinant.

1. $\begin{vmatrix} 1 & 1 \\ -7 & 1 \end{vmatrix}$

2. $\begin{vmatrix} -2 & 1 \\ -2 & 4 \end{vmatrix}$

3. $\begin{vmatrix} 1 & 1 \\ 2 & 3 \end{vmatrix}$

4. $\begin{vmatrix} 3 & 1 \\ 2 & -2 \end{vmatrix}$

5. $\begin{vmatrix} 2 & 4 & -12 \\ 3 & 2 & 1 \\ 3 & 1 & 7 \end{vmatrix}$

6. $\begin{vmatrix} 2 & -5 & -1 \\ 1 & -1 & 6 \\ 2 & -3 & 9 \end{vmatrix}$

7. $\begin{vmatrix} 2 & 3 & 11 \\ 3 & -1 & 4 \\ 2 & -1 & 1 \end{vmatrix}$

8. $\begin{vmatrix} 8 & -4 & 3 \\ 2 & -5 & 6 \\ 6 & 2 & 1 \end{vmatrix}$

Solve each equation for *x*.

9. $\begin{vmatrix} x & 10 \\ 2 & 3 \end{vmatrix} = 1$

10. $\begin{vmatrix} 3 & 4 \\ 5 & x \end{vmatrix} = 3$

11. $\begin{vmatrix} 1 & 2x & -4 \\ 1 & x & 3 \\ -2 & -1 & -1 \end{vmatrix} = 0$

12. $\begin{vmatrix} x & -2 & 1 \\ 3 & 2 & 2 \\ 3 & -2 & 1 \end{vmatrix} = 0$

Find the area of each triangle having vertices at *P*, *Q*, and *R*.

13. $P(2,-3)$, $Q(-3,-4)$, $R(5,6)$

14. $P(2,3)$, $Q(3,-2)$, $R(-2,4)$

Evaluate each determinant.

15. $\begin{vmatrix} 1 & 2 & -1 & 2 \\ 3 & 0 & 2 & 4 \\ 1 & 0 & -1 & -2 \\ 0 & 3 & 2 & -1 \end{vmatrix}$

16. $\begin{vmatrix} 1 & 0 & 0 & 1 \\ 2 & -1 & 0 & 2 \\ 1 & 2 & 3 & -1 \\ 1 & 2 & -1 & 1 \end{vmatrix}$

Use Cramer's Rule to solve each system of equations.

17. $2x + y = 8$
$3x + 5y = 10$

18. $3x + 5y = 4$
$5x + 2y = 9$

19. $x - 3y - 4z = 12$
$x - 3y + 2z = 0$
$x - y + 2z = 6$

20. $2x - 5y - 4z = 3$
$x - 6y + 11z = -6$
$x - 2y - z = 2$

Find the partial fraction decomposition for each rational expression.

1. $\dfrac{6x+2}{5x(x-3)}$

2. $\dfrac{x-2}{(x-5)^2}$

3. $\dfrac{5x^2+30x+27}{3x(x+3)^2}$

4. $\dfrac{3x^2+x+2}{(x+2)(x+3)(x-1)}$

5. $\dfrac{-x-4}{x(x+1)}$

6. $\dfrac{-x-8}{x(x+4)}$

7. $\dfrac{8x-5}{x(x-1)}$

8. $\dfrac{3-3x}{(x+1)(2x-1)}$

9. $\dfrac{2x^2+20x+26}{(x-1)^2(x+3)^2}$

10. $\dfrac{-2x^2+28x-6}{x(x-3)(x+2)}$

11. $\dfrac{5x-13}{x^2-5x+6}$

12. $\dfrac{9x^2-9x+6}{(2x-1)(x-1)(x+2)}$

13. $\dfrac{2x-5}{x^2-6x+9}$

14. $\dfrac{2x^2+10x+2}{(x+1)(x-2)(x+3)}$

15. $\dfrac{x+12}{(2x-1)(x+2)}$

16. $\dfrac{-7x-14}{3x^2-10x-8}$

17. $\dfrac{18x+16}{3x^2+8x+4}$

18. $\dfrac{6x^2+22x+18}{(x+1)(x+2)(x+3)}$

19. $\dfrac{3x^2+10x+12}{(x+1)^3}$

20. $\dfrac{5x^2-29x+21}{(2x+1)(x-3)^2}$

A non-linear system is given below. Verify that the given points are solutions of the system.

1. $y = x^2 - 4x + 5$; $3x - 2y = 4$; $\left(\frac{7}{2}, \frac{13}{4}\right)$ and $(2,1)$

2. $y = 6x^2 - x$; $y = x$; $(0,0)$ and $\left(\frac{1}{3}, \frac{1}{3}\right)$

3. $x^2 + y^2 = 4$; $3x + y = 2$; $\left(\frac{6}{5}, -\frac{8}{5}\right)$ and $(0,2)$

4. $y = \ln(x+3)$; $2x + y = 3$; $(0.828733, 1.34253)$

Give all solutions of each nonlinear system of equations.

5. $2x + 3y = 12$
   $4x^2 + 9y^2 = 144$

6. $9x^2 + 16y^2 = 144$
   $9x^2 + 5y^2 = 45$

7. $xy = 4$
   $2x^2 + y^2 = 18$

8. $9x^2 + 16y^2 = 144$
   $9x^2 - 16y^2 = 144$

9. $x^2 + y^2 = 9$
   $4x - 3y = 0$

10. $y = x^2 + 3x + 2$
    $y = 2x + 3$

11. $x^2 + y^2 = 1$
    $y = x - 1$

12. $y = x^2 + 1$
    $2x^2 + 2y = 6$

13. $y = x^2$
    $y = 8 - x^2$

14. $9x^2 - 16y^2 = 144$
    $y^2 - 4x^2 = 1$

15. $xy = 2$
    $x + y = 4$

16. $3x - y = -2$
    $2x^2 - y = 0$

17. $xy = 4$
    $x^2 + y^2 = 8$

18. $x^2 = y + 8$
    $x^2 + y^2 = 64$

19. $y = x + 4$
    $x^2 + y^2 = 16$

20. $x^2 - 2y = 8$
    $x^2 + y^2 = 16$

21. $x^2 + y^2 = 4$
    $-x^2 + y = 8$

Graph each inequality.

1. $2x + 3y > 6$
2. $x + y \leq 5$
3. $2x \geq y + 8$
4. $4y - 3x \leq 12$
5. $x + y \geq 8$
6. $3y \leq 2x + 12$
7. $2x + 2y > 4$
8. $y - 4x \geq -10$
9. $y < 4x^2 - 6$
10. $(x-2)^2 + (y+1)^2 < 4$

Graph the solution set of each system of inequalities.

11. $y < 9 - 3x$
    $x > -3y + 1$
12. $x < 2y - 5$
    $x + 4y \leq -6$
13. $y \geq -x$
    $y \geq -3$
14. $x + y \leq 8$
    $x - y \geq 2$
15. $y < 16 - x^2$
    $y > x^2 - 4$
16. $y \geq 2$
    $x - y > 0$
17. $y \geq 0$
    $x + 3y \geq 6$
18. $y \geq e^x$
    $y \geq 4$
19. $y \leq \left(\frac{1}{3}\right)^x$
    $y \geq 3$
20. $y \leq \log x$
    $y \geq x - 2$

Find the values of the variables for which each statement is true, if possible.

1. $\begin{bmatrix} 3 & x \\ y & 4 \end{bmatrix} = \begin{bmatrix} 3 & -2 \\ -1 & 4 \end{bmatrix}$

2. $\begin{bmatrix} 3 & x \\ y & -3 \end{bmatrix} = \begin{bmatrix} z & -2 \\ 4 & w \end{bmatrix}$

3. $\begin{bmatrix} 1 & 1 & 1 \\ 3 & 2 & -1 \\ 2 & 3 & 4 \end{bmatrix} = \begin{bmatrix} a & 1 & 1 \\ b & c & -1 \\ 2 & 3 & d \end{bmatrix}$

4. $\begin{bmatrix} x-2 & 1 & 1 \\ 2 & 3 & 3 \\ 4 & 5 & 2 \end{bmatrix} = \begin{bmatrix} 4 & -1 & z-1 \\ y+3 & 3 & 3 \\ 4 & 5 & 2 \end{bmatrix}$

5. $[x \quad y \quad z] = [6 \quad 3]$

Perform each operation, if possible.

6. $\begin{bmatrix} 4 & 3 \\ 3 & 2 \end{bmatrix} + \begin{bmatrix} 4 & 1 \\ 2 & 3 \end{bmatrix}$

7. $\begin{bmatrix} 5 & 3 \\ 2 & 3 \end{bmatrix} - \begin{bmatrix} 3 & 0 \\ 1 & 2 \end{bmatrix}$

8. $\begin{bmatrix} 1 & 3 \\ 4 & 2 \end{bmatrix} + \begin{bmatrix} 2 & -1 \\ 3 & 0 \end{bmatrix}$

9. $\begin{bmatrix} 2 & 1 & 0 \\ -1 & 0 & 2 \\ 3 & -1 & 3 \end{bmatrix} + \begin{bmatrix} 2 & 0 & -1 \\ 1 & 2 & 3 \\ -1 & 2 & 1 \end{bmatrix}$

10. $\begin{bmatrix} 2 & 0 & 4 \\ -1 & 1 & -2 \\ 3 & 0 & 2 \end{bmatrix} - \begin{bmatrix} -1 & 2 & 1 \\ 2 & 1 & 3 \\ 5 & 2 & 0 \end{bmatrix}$

11. $\begin{bmatrix} -x & 3y & 2x \\ 1+x & 2x & 0 \\ x-2 & y-1 & -3y \end{bmatrix} - \begin{bmatrix} x & x & x \\ -1 & 2y & -3 \\ x+2 & 2y & 5y \end{bmatrix}$

Let $A = \begin{bmatrix} 2 & 3 \\ 1 & -1 \end{bmatrix}$ **and** $B = \begin{bmatrix} 4 & 5 \\ 0 & 7 \end{bmatrix}$. **Find each of the following.**

12. $3A$

13. $\frac{1}{2}B$

14. $A-2B$

15. $3A+2B$

Find each matrix product.

16. $\begin{bmatrix} 1 & 2 \\ 2 & -1 \end{bmatrix} \begin{bmatrix} 3 & 4 \\ 1 & -3 \end{bmatrix}$

17. $\begin{bmatrix} 1 & 2 & 3 \\ 0 & 0 & 1 \end{bmatrix} \begin{bmatrix} 0 & 1 \\ -1 & 2 \\ 2 & 1 \end{bmatrix}$

18. $\begin{bmatrix} 5 & -1 & 11 \\ -3 & -2 & -2 \\ 2 & 4 & 1 \end{bmatrix} \begin{bmatrix} 2 & 1 & 3 \\ 3 & -1 & 2 \\ 4 & 1 & 2 \end{bmatrix}$

19. $\begin{bmatrix} 0 & 1 & 0 \\ -1 & 4 & 3 \\ 2 & 1 & 0 \end{bmatrix} \begin{bmatrix} 3 & -1 \\ 2 & 0 \\ -1 & 1 \end{bmatrix}$

Decide whether or not the given matrices are inverses of each other.

1. $\begin{bmatrix} 2 & -3 \\ 4 & -7 \end{bmatrix}$ and $\begin{bmatrix} 3.5 & -1.5 \\ 2 & -1 \end{bmatrix}$

2. $\begin{bmatrix} 1 & 1 \\ 3 & 3 \end{bmatrix}$ and $\begin{bmatrix} 7 & -2 \\ -6 & 2 \end{bmatrix}$

3. $\begin{bmatrix} 1 & 5 & 2 \\ 1 & 1 & 7 \\ 0 & -3 & 4 \end{bmatrix}$ and $\begin{bmatrix} -25 & 26 & -33 \\ 4 & -4 & 5 \\ 3 & -3 & 4 \end{bmatrix}$

4. $\begin{bmatrix} 3 & 0 & 0 \\ 3 & 2 & 0 \\ 3 & 2 & 5 \end{bmatrix}$ and $\begin{bmatrix} 1/3 & 0 & 0 \\ -1/2 & 1/2 & 0 \\ 0 & -1/5 & 1/5 \end{bmatrix}$

Find the inverse for each matrix.

5. $\begin{bmatrix} 5 & -2 \\ 3 & 7 \end{bmatrix}$

6. $\begin{bmatrix} -7 & 3 \\ 5 & -2 \end{bmatrix}$

7. $\begin{bmatrix} 1 & -1 & 3 \\ 2 & 1 & 2 \\ -2 & -2 & 1 \end{bmatrix}$

8. $\begin{bmatrix} 0 & 5 & 3 \\ 1 & 2 & 1 \\ 0 & 3 & 2 \end{bmatrix}$

Solve each system by using the inverse of the coefficient matrix.

9. $x + y = 4$
 $2x + 3y = 9$

10. $2x + y = 7$
 $x - y = -4$

11. $7x + 8y = 20$
 $3x + 5y = 12$

12. $x + 2y = 3$
 $2x + y = 1$

13. $x+2y-3z = -4$
 $2x - y + z = 3$
 $3x + 2y + z = 10$

14. $x - y = 4$
 $x - z = 0$
 $-6x + 2y + 3z = 7$

Use a graphing calculator to find the inverse of each matrix.

15. $\begin{bmatrix} \sqrt{7} & -2 \\ 1/4 & \sqrt{3} \end{bmatrix}$

16. $\begin{bmatrix} 0.1 & -0.9 & 0.3 \\ 0.2 & 0.4 & 0.2 \\ 0.3 & -0.6 & 0.9 \end{bmatrix}$

Use a graphing calculator and the method of matrix inverses to solve each system.

17. $3.1x + \sqrt{2}y = 6$
 $\sqrt{3}x - 6y = 4$

18. $\sqrt{3}x + \pi y + ez = 2$
 $ex + \pi y + \sqrt{3}z = 1$
 $\pi x + \sqrt{3}y + ez = 3$

5.1 Systems of Linear Equations

1. $(2,14)$

2. $\left(\frac{10}{3},-\frac{4}{3}\right)$

3. $(3,1)$

4. $\left(\frac{15}{4},-\frac{9}{4}\right)$

5. $\left(\frac{7}{2},-\frac{19}{4}\right)$

6. $(2,1)$

7. $(9,3)$

8. $\left(\frac{23}{11},\frac{25}{11}\right)$

9. $\varnothing$, inconsistent

10. $\left(\frac{5y+6}{2},y\right)$, infinitely many solutions

11. $\left(\frac{1-2y}{6},y\right)$, infinitely many solutions

12. $\varnothing$, inconsistent

13. $(1.794,1.587)$

14. $(2.646,-1.315)$

15. $(0.620,-1.283)$

16. $(1.439,-0.883)$

17. $(2,-1,3)$

18. $\left(\frac{109}{32},\frac{87}{32},-\frac{1}{8}\right)$

5.2 Matrix Solution of Linear Systems

1. $\left[\begin{array}{cc|c} 5 & -2 & -10 \\ 4 & 3 & -1 \end{array}\right]$

2. $\left[\begin{array}{ccc|c} 3 & 6 & -3 & -1 \\ -4 & -3 & -4 & 3 \\ 2 & 2 & 4 & 2 \end{array}\right]$

3. $\begin{aligned} x+y-z&=2 \\ 2x-y+z&=6 \\ -x+2y-3z&=1 \end{aligned}$

4. $\begin{aligned} 3x-4y+8z&=0 \\ 5x+3y-4z&=-2 \\ -5x+3y-8z&=-3 \end{aligned}$

5. $\left(\frac{27}{5},\frac{18}{5}\right)$

6. $\left(-\frac{5}{2},-2\right)$

7. $\left(-22,\frac{31}{2}\right)$

8. $\left(\frac{9}{17},-\frac{28}{17}\right)$

9. $\left(\frac{65}{162},\frac{35}{162},\frac{2}{27}\right)$

10. $\left(\frac{13}{7},-\frac{39}{14},\frac{47}{14}\right)$

11. $\left(\frac{1}{3},\frac{20}{3},-\frac{11}{3}\right)$

12. $\varnothing$

13. $\left(6,\frac{31}{5},-\frac{13}{5}\right)$

14. $\left(-\frac{19}{20},\frac{23}{20},\frac{3}{4}\right)$

15. $\left(\frac{2}{5},\frac{7}{10},-\frac{8}{5}\right)$

16. $\left(\frac{3}{8}, -\frac{7}{8}, \frac{1}{8}\right)$

17. $(-1, 6, -1)$

18. $\left(1, -\frac{2}{3}, 2\right)$

5.3 Determinant Solution of Linear Systems

1. 8

2. −6

3. 1

4. −8

5. −10

6. 4

7. 10

8. −170

9. 7

10. $\frac{23}{3}$

11. $\frac{7}{19}$

12. 3

13. 21

14. 9.5

15. −110

16. 2

17. $\left(\frac{30}{7}, -\frac{4}{7}\right)$

18. $\left(\frac{37}{19}, -\frac{7}{19}\right)$

19. $(13, 3, -2)$

20. $\left(\frac{23}{5}, \frac{7}{5}, -\frac{1}{5}\right)$

5.4 Partial Fractions

1. $-\frac{2}{15x} + \frac{4}{3(x-3)}$

2. $\frac{1}{x-5} + \frac{3}{(x-5)^2}$

3. $\frac{1}{x} + \frac{2}{3(x+3)} + \frac{2}{(x+3)^2}$

4. $\frac{-4}{x+2} + \frac{13}{2(x+3)} + \frac{1}{2(x-1)}$

5. $-\frac{4}{x} + \frac{3}{x+1}$

6. $-\frac{2}{x} + \frac{1}{x+4}$

7. $\frac{5}{x} + \frac{3}{x-1}$

8. $-\frac{2}{x+1} + \frac{1}{2x-1}$

9. $\frac{3}{(x-1)^2} - \frac{1}{(x+3)^2}$

10. $\frac{1}{x} + \frac{4}{(x-3)} - \frac{7}{(x+2)}$

11. $\frac{3}{x-2} + \frac{2}{x-3}$

12. $\frac{2}{x-1} - \frac{3}{2x-1} + \frac{4}{x+2}$

13. $\frac{2}{x-3} + \frac{1}{(x-3)^2}$

14. $\frac{1}{x+1} + \frac{2}{x-2} - \frac{1}{x+3}$

15. $\frac{5}{2x-1} - \frac{2}{x+2}$

16. $\frac{2}{3x+2} - \frac{3}{x-4}$

17. $\frac{3}{3x+2} + \frac{5}{x+2}$

18. $\frac{1}{x+1} + \frac{2}{x+2} + \frac{3}{x+3}$

19. $\frac{3}{x+1} + \frac{4}{(x+1)^2} + \frac{5}{(x+1)^3}$

20. $\frac{3}{2x+1} + \frac{1}{x-3} - \frac{3}{(x-3)^2}$

5.5 Nonlinear Systems of Equations

1. yes

2. yes

3. yes

4. yes

5. $(6,0);\ (0,4)$

6. $(0,3),\ (0,-3)$

7. $\left(2\sqrt{2},\sqrt{2}\right);\ \left(-2\sqrt{2},-\sqrt{2}\right);\ \left(\sqrt{2},2\sqrt{2}\right);\ \left(-\sqrt{2},-2\sqrt{2}\right)$

8. $(4,0);\ (-4,0)$

9. $\left(\frac{9}{5},\frac{12}{5}\right);\ \left(-\frac{9}{5},-\frac{12}{5}\right)$

10. $\left(\frac{-1-\sqrt{5}}{2},2-\sqrt{5}\right);\ \left(\frac{-1+\sqrt{5}}{2},2+\sqrt{5}\right)$

11. $(1,0);\ (0,-1)$

12. $(1,2);\ (-1,2)$

13. $(2,4);\ (-2,4)$

14. $\varnothing$

15. $\left(2-\sqrt{2},2+\sqrt{2}\right);\ \left(2+\sqrt{2},2-\sqrt{2}\right)$

16. $\left(-\frac{1}{2},\frac{1}{2}\right);\ (2,8)$

17. $(2,2);\ (-2,-2)$

18. $(-\sqrt{15},7);(\sqrt{15},7);(0,-8)$

19. $(0,4);\ (-4,0)$

20. $(0,-4);\ (2\sqrt{3},2);\ (-2\sqrt{3},2)$

21. $\varnothing$

5.6 Systems of Inequalities and Linear Programming

1.

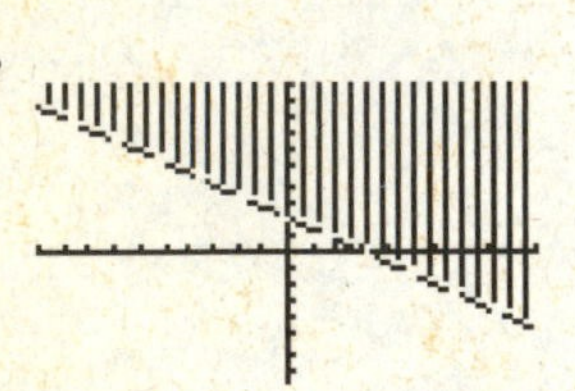

2.

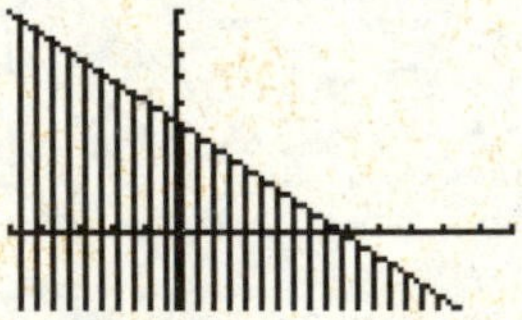

3.

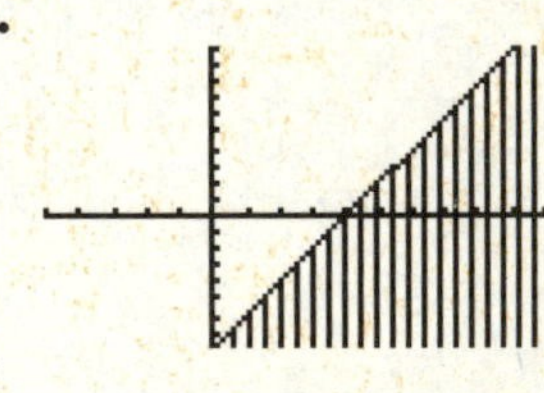

4.

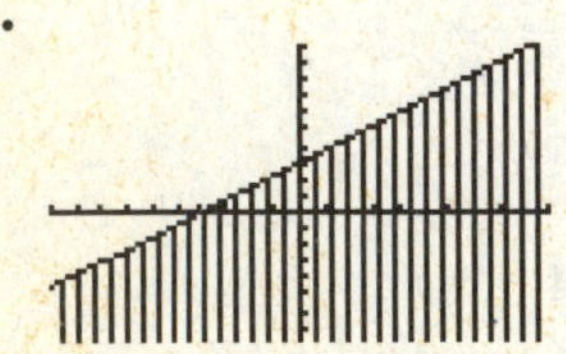

5.

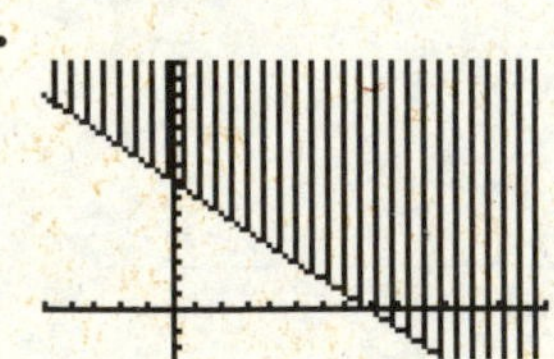

6.

7.
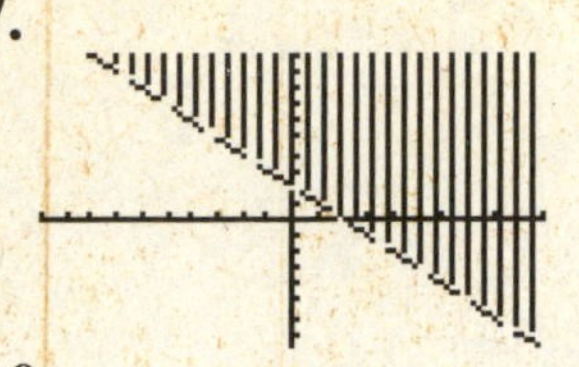

8.
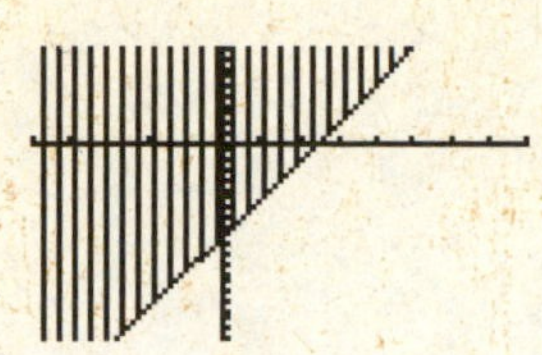

9.
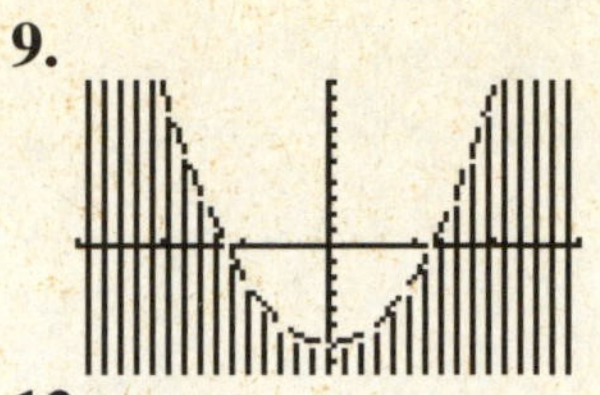

10.
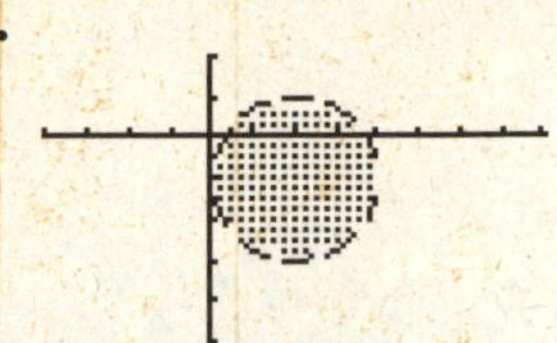

11.
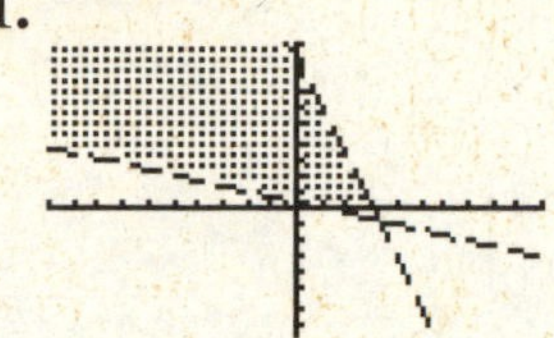

12.
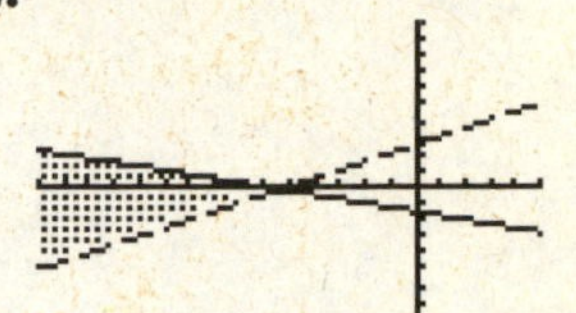

13.
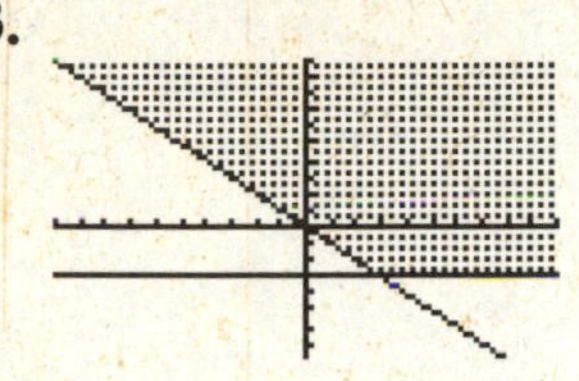

14.
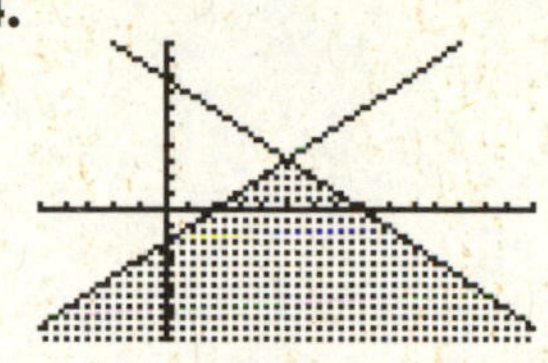

15.
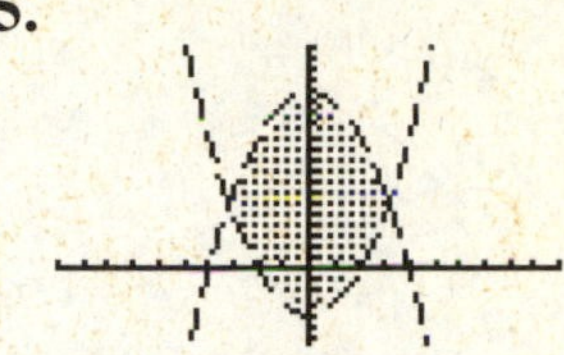

16.
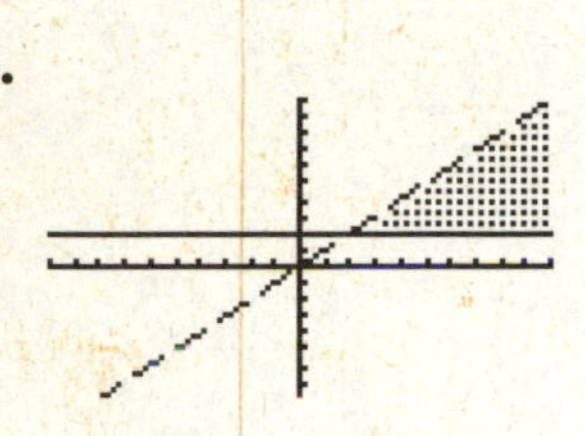

17.
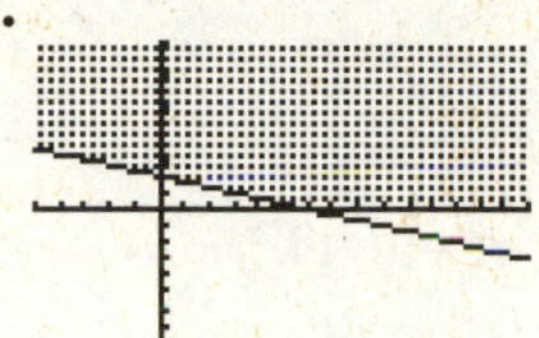

18.h
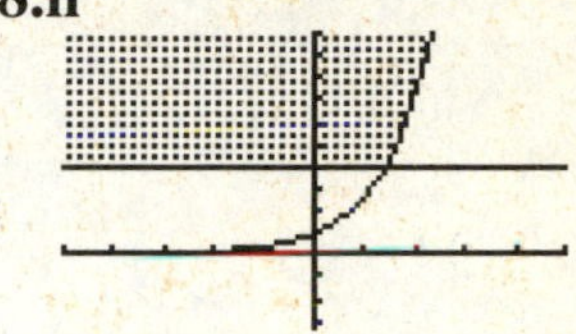

19.
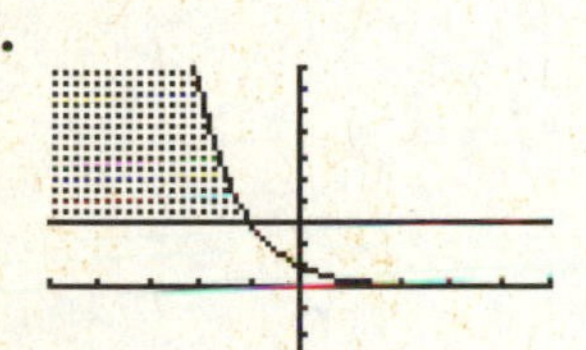

20.
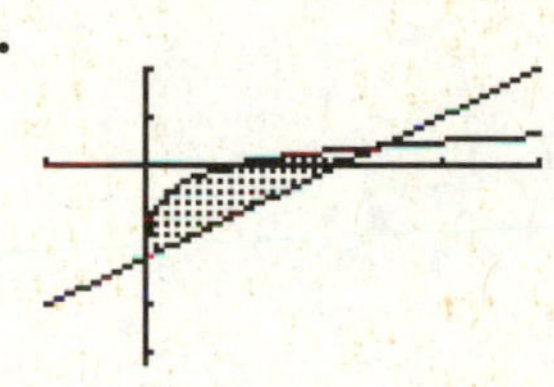

5.7 Properties of Matrices

1. $x = -2, y = -1$

2. $x = -2, y = 4, w = -3, z = 3$

3. $a = 1, b = 3, c = 2, d = 4$

4. $x = 6, y = -1, z = 2$

5. not possible

6. $\begin{bmatrix} 8 & 4 \\ 5 & 5 \end{bmatrix}$

7. $\begin{bmatrix} 2 & 3 \\ 1 & 1 \end{bmatrix}$

8. $\begin{bmatrix} 3 & 2 \\ 7 & 2 \end{bmatrix}$

9. $\begin{bmatrix} 4 & 1 & -1 \\ 0 & 2 & 5 \\ 2 & 1 & 4 \end{bmatrix}$

10. $\begin{bmatrix} 3 & -2 & 3 \\ -3 & 0 & -5 \\ -2 & -2 & 2 \end{bmatrix}$

11. $\begin{bmatrix} -2x & 3y-x & x \\ 2+x & 2x-2y & 3 \\ -4 & -y-1 & -8y \end{bmatrix}$

12. $\begin{bmatrix} 6 & 9 \\ 3 & -3 \end{bmatrix}$

13. $\begin{bmatrix} 2 & 5/2 \\ 0 & 7/2 \end{bmatrix}$

14. $\begin{bmatrix} -6 & -7 \\ 1 & -15 \end{bmatrix}$

15. $\begin{bmatrix} 14 & 19 \\ 3 & 11 \end{bmatrix}$

16. $\begin{bmatrix} 5 & -2 \\ 5 & 11 \end{bmatrix}$

17. $\begin{bmatrix} 4 & 8 \\ 2 & 1 \end{bmatrix}$

18. $\begin{bmatrix} 51 & 17 & 35 \\ -20 & -3 & -17 \\ 20 & -1 & 16 \end{bmatrix}$

19. $\begin{bmatrix} 2 & 0 \\ 2 & 4 \\ 8 & -2 \end{bmatrix}$

5.8 Matrix Inverses

1. yes

2. no

3. yes

4. yes

5. $\begin{bmatrix} 7/41 & 2/41 \\ -3/41 & 5/41 \end{bmatrix}$

6. $\begin{bmatrix} 2 & 3 \\ 5 & 7 \end{bmatrix}$

7. $\begin{bmatrix} 1 & -1 & -1 \\ -6/5 & 7/5 & 4/5 \\ -2/5 & 4/5 & 3/5 \end{bmatrix}$

8. $\begin{bmatrix} -1 & 1 & 1 \\ 2 & 0 & -3 \\ -3 & 0 & 5 \end{bmatrix}$

9. $(3,1)$

10. $(1,5)$

11. $\left(\frac{4}{11}, \frac{24}{11}\right)$

12. $\left(-\frac{1}{3}, \frac{5}{3}\right)$

13. $(1,2,3)$

14. $(-15,-19,-15)$

15. $\begin{bmatrix} 0.341 & 0.394 \\ -0.049 & 0.521 \end{bmatrix}$

16. $\begin{bmatrix} 5.714 & 7.5 & -3.571 \\ -1.429 & 0 & 0.476 \\ -2.857 & -2.5 & 2.619 \end{bmatrix}$

17. $(1.979, -0.095)$

18. $(0.194, -0.515, 1.208)$

Graph each horizontal parabola.

1. $y^2 = 4x$
2. $2y^2 = -10x$
3. $2x = y^2 - 8y + 12$
4. $x = y^2 + y - 2$
5. $x + 5 = -(y-1)^2$
6. $3(x-2) = (y+4)^2$

Give the focus, directrix, and axis for each parabola.

7. $y^2 + 10x = 0$
8. $8x = y^2 - 2y - 15$
9. $x^2 - 4x = -2y - 6$
10. $y = 3(x+1)^2 - 2$
11. $x^2 + 6x + 8y + 25 = 0$
12. $-10(y-3) = (x+2)^2$

Write an equation for each parabola with vertex at the origin.

13. focus: $(4,0)$
14. focus: $(2,0)$
15. focus: $\left(0,-\frac{3}{2}\right)$
16. focus: $(0,2)$
17. through $(2,4)$, opening up
18. through $(2,4)$, opening to the right

Write an equation for each parabola.

19. vertex: $(-2,3)$, focus: $(-2,5)$
20. vertex: $(5,4)$, focus: $(2,4)$

Determine the two equations necessary to graph each horizontal parabola using a graphing calculator.

21. $(y-2)^2 = 8(x+1)$
22. $x = \frac{1}{4}(y-2)^2 - 1$

Identify the domain, range, center, vertices, endpoints of the minor axis, and the foci in each figure.

1. $\frac{x^2}{9}+\frac{y^2}{4}=1$

2. $9x^2+4y^2=36$

3. $16x^2+25y^2=400$

4. $\frac{(x+2)^2}{9}+\frac{(y+2)^2}{4}=1$

5. $(x-3)^2+\frac{(y-5)^2}{9}=1$

6. $9x^2+16y^2=144$

7. $\frac{(x-1)^2}{9}+\frac{(y+1)^2}{4}=1$

8. $y^2+36x^2=36$

Write an equation for each ellipse.

9. center: $(0,0)$; focus: $(3,0)$; vertex: (5,0)

10. foci: $(0,2),(0,-2)$; vertices: $(0,5),(0,-5)$

11. x-intercepts: $(4,0),(-4,0)$ y-intercepts: $(0,3),(0,-3)$

12. x-intercepts: $(3,0),(-3,0)$ y-intercepts: $(0,4),(0,-4)$

13. foci: $(3,0),(-3,0)$; vertices: $(-7,0),(7,0)$

14. vertices: $(0,-2),(0,2)$ length of minor axis = 2

15. vertices: $(0,-8),(0,8)$ length of minor axis = 10

16. center: (0,0); length of horizontal major axis = 8; length of minor axis = 4

Graph each equation. Give the domain and range. Identify any that are graphs of functions.

17. $x=\sqrt{1-\frac{y^2}{16}}$

18. $y=-\sqrt{1-\frac{x^2}{9}}$

Determine the two equations necessary to graph each ellipse with a graphing calculator.

19. $\frac{x^2}{4}+y^2=1$

20. $\frac{x^2}{6}+\frac{y^2}{10}=1$

Find the eccentricity of each ellipse.

21. $\frac{x^2}{5}+\frac{y^2}{9}=1$

22. $3x^2+4y^2=12$

Give the domain, range, center, vertices, foci, and equations of the asymptotes.

1. $\frac{x^2}{25} - \frac{y^2}{9} = 1$

2. $\frac{(x-1)^2}{4} - \frac{(y-3)^2}{4} = 1$

3. $x^2 - y^2 = 1$

4. $25y^2 - 4x^2 = 100$

5. $\frac{(y-1)^2}{4} - \frac{(x+3)^2}{16} = 1$

6. $\frac{y^2}{9} - x^2 = 1$

7. $2x^2 - y^2 = 4$

Graph each equation. Give the domain and range. Identify any that are the graphs of functions.

8. $y = \sqrt{\frac{x^2}{9} - 1}$

9. $x = -\sqrt{1 + \frac{y^2}{9}}$

Find the eccentricity of each hyperbola.

10. $y^2 - \frac{x^2}{4} = 1$

11. $\frac{x^2}{8} - \frac{y^2}{4} = 1$

Write an equation for each hyperbola.

12. center: (1,4); focus: $(-2,4)$; vertex: (0,4)

13. vertices: $(2,-1), (10,-1)$; foci: $(0,-1), (12,-1)$

14. vertices: $(3,0), (-3,0)$; foci: $(5,0), (-5,0)$

15. y-intercepts: $(0,4), (0,-4)$; foci: $(0,5), (0,-5)$

16. vertices: $(0,3), (0,-3)$; foci: $(0,5), (0,-5)$

17. vertices: $(2,0), (6,0)$; foci: $(0,0), (8,0)$

18. center: $(-3,1)$; focus: $(-3,6)$; vertex: $(-3,4)$

Determine the two equations necessary to graph each hyperbola with a graphing calculator.

19. $3y^2 - 2x^2 = 12$

20. $3x^2 - y^2 = 6x$

The equation of a conic section is given in a familiar form. Identify the type of graph that each equation has.

1. $\frac{y^2}{25}+\frac{x^2}{16}=1$
2. $y^2-x^2+2x=2$
3. $\frac{x^2}{4}-\frac{y^2}{16}=1$
4. $y^2=5x$
5. $y^2+x^2=9$
6. $4y^2-9x^2=36$
7. $2x^2+6y^2=12$
8. $\frac{x^2}{3}+\frac{y^2}{2}=1$

For each equation that has a graph, identify the type of graph.

9. $4x^2+y^2-16x-6y=15$
10. $x^2-8x+6y=0$
11. $y^2-8y-8x^2+4x=8$
12. $x^2-8x=4y$
13. $x^2-2x-y^2-2y=1$
14. $x^2+16y=0$
15. $5x^2+3y^2=15$
16. $x^2+2x-4y^2=3$
17. $6y^2-24x=0$
18. $2y^2+8y=9x^2$
19. $3x^2+4y^2=12$
20. $4x^2-6y^2-8x+24y-44=0$

6.1 Parabolas

1.

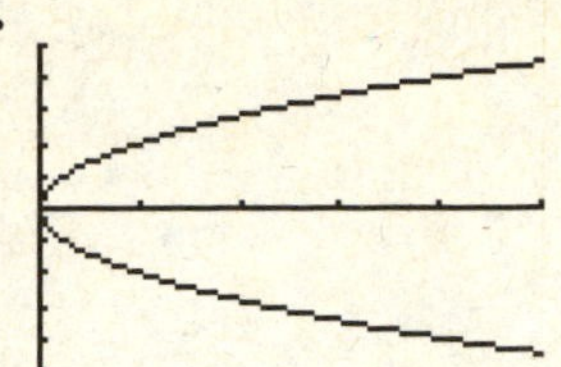

2.

3.

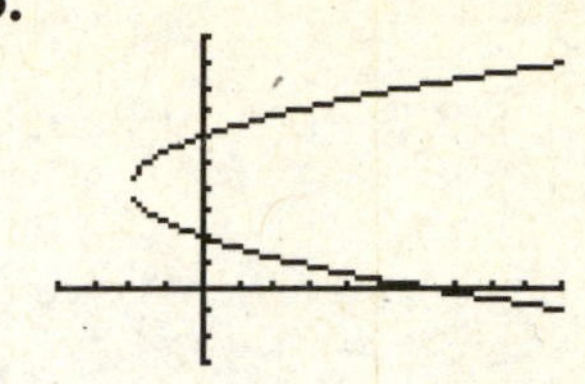

4.

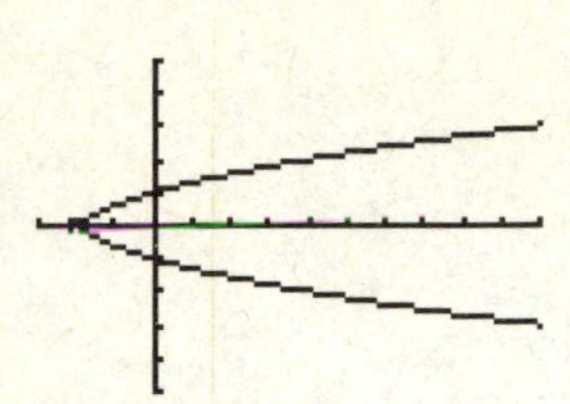

5.

6.

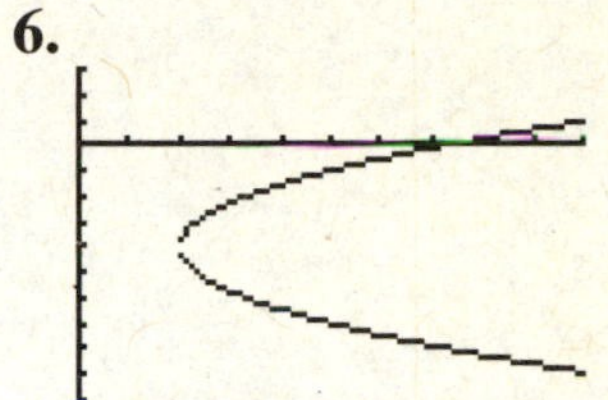

7. $\left(-\frac{5}{2},0\right)$; $x=\frac{5}{2}$; $y=0$

8. $(0,1)$; $x=-4$; $y=1$

9. $\left(2,-\frac{3}{2}\right)$; $y=-\frac{1}{2}$; $x=2$

10. $\left(-1,-\frac{23}{12}\right)$; $y=-\frac{25}{12}$; $x=-1$

11. $(-3,-4)$; $y=0$; $x=-3$

12. $\left(-2,\frac{1}{2}\right)$; $y=\frac{11}{2}$; $x=-2$

13. $y^2=16x$

14. $y^2=8x$

15. $x^2=-6y$

16. $x^2=8y$

17. $x^2=y$

18. $y^2=8x$

19. $(x+2)^2=8(y-3)$

20. $(y-4)^2=-12(x-5)$

21. $y=2+2\sqrt{2(x+1)}$; $y=2-2\sqrt{2(x+1)}$

22. $y=2+2\sqrt{x+1}$ $y=2-2\sqrt{x+1}$

6.2 Ellipses

1. $[-3,3]$; $[-2,2]$; $(0,0)$; $(3,0),(-3,0)$; $(0,2),(0,-2)$; $(\sqrt{5},0),(-\sqrt{5},0)$

2. $[-2,2]$; $[-3,3]$; $(0,0)$; $(0,3),(0,-3)$; $(2,0),(-2,0)$; $(0,\sqrt{5}),(0,-\sqrt{5})$

3. $[-5,5]$; $[-4,4]$; $(0,0)$; $(5,0),(-5,0)$; $(0,4),(0,-4)$; $(3,0),(-3,0)$

4. $[-5,1]$; $[-4,0]$; $(-2,-2)$; $(-5,-2),(1,-2)$; $(-2,0),(-2,-4)$; $(-2+\sqrt{5},-2),(-2-\sqrt{5},-2)$

5. $[2,4]$; $[2,8]$; $(3,5)$; $(3,8),(3,2)$; $(2,5),(4,5)$; $(3,5+2\sqrt{2}),(3,5-2\sqrt{2})$

6. $[-4,4]$; $[-3,3]$; $(0,0)$; $(4,0),(-4,0)$; $(0,3),(0,-3)$; $(\sqrt{7},0),(-\sqrt{7},0)$

7. $[-2,4]$; $[-3,1]$; $(1,-1)$; $(-2,-1),(4,-1)$; $(1,1),(1,-3)$; $(1+\sqrt{5},-1),(1-\sqrt{5},-1)$

8. $[-1,1]$; $[-6,6]$; $(0,0)$; $(0,6),(0,-6)$; $(1,0),(-1,0)$; $(0,\sqrt{35}),(0,-\sqrt{35})$

9. $\frac{x^2}{25}+\frac{y^2}{16}=1$

10. $\frac{x^2}{21}+\frac{y^2}{25}=1$

11. $\frac{x^2}{16}+\frac{y^2}{9}=1$

12. $\frac{x^2}{9}+\frac{y^2}{16}=1$

13. $\frac{x^2}{49}+\frac{y^2}{40}=1$

14. $x^2+\frac{y^2}{4}=1$

15. $\frac{x^2}{25}+\frac{y^2}{64}=1$

16. $\frac{x^2}{16}+\frac{y^2}{4}=1$

17. $[0,1]$; $[-4,4]$

18. $[-3,3]$; $[-1,0]$; function

19. $y=\sqrt{1-\frac{x^2}{4}}$; $y=-\sqrt{1-\frac{x^2}{4}}$

20. $y=\sqrt{10\left(1-\frac{x^2}{6}\right)}$; $y=-\sqrt{10\left(1-\frac{x^2}{6}\right)}$

21. $\frac{2}{3}$

22. $\frac{1}{2}$

6.3 Hyperbolas

1. $(-\infty,-5]\cup[5,\infty)$; $(-\infty,\infty)$; $(0,0)$; $(5,0),(-5,0)$; $(\sqrt{34},0),(-\sqrt{34},0)$; $y=\pm\frac{3}{5}x$

2. $(-\infty,-1]\cup[3,\infty)$; $(-\infty,\infty)$; $(1,3)$; $(-1,3),(3,3)$; $(1+2\sqrt{2},3),(1-2\sqrt{2},3)$; $y=\pm x$

3. $(-\infty,-1]\cup[1,\infty)$; $(-\infty,\infty)$; $(0,0)$; $(1,0),(-1,0)$; $(\sqrt{2},0),(-\sqrt{2},0)$; $y=\pm x$

4. $(-\infty,\infty)$; $(-\infty,-2]\cup[2,\infty)$; $(0,0)$; $(0,2),(0,-2)$; $(0,\sqrt{29}),(0,-\sqrt{29})$; $y=\pm\frac{2}{5}x$

5. $(-\infty,\infty)$; $(-\infty,-1]\cup[3,\infty)$; $(-3,1)$; $(-3,3),(-3,-1)$; $(-3,1+2\sqrt{5}),(-3,1-2\sqrt{5})$; $y-1=\pm\frac{1}{2}(x+3)$

6. $(-\infty,\infty);(-\infty,-3]\cup[3,\infty);\ (0,0);\ (0,3),(0,-3);\ (0,\sqrt{10}),(0,-\sqrt{10});\ y=\pm\,3x$

7. $(-\infty,-\sqrt{2}]\cup[\sqrt{2},\infty);\ (-\infty,\infty);\ (0,0);\ (\sqrt{2},0),(-\sqrt{2},0);\ (\sqrt{6},0),(-\sqrt{6},0);\ y=\pm\sqrt{2}x$

8. $(-\infty,-3]\cup[3,\infty);\ [0,\infty);$ function

9. $(-\infty,-1];\ (-\infty,\infty)$

10. $\sqrt{5}$

11. $\frac{\sqrt{6}}{2}$

12. $(x-1)^2-\frac{(y-4)^2}{8}=1$

13. $\frac{(x-6)^2}{16}-\frac{(y+1)^2}{20}=1$

14. $\frac{x^2}{9}-\frac{y^2}{16}=1$

15. $\frac{y^2}{16}-\frac{x^2}{9}=1$

16. $\frac{y^2}{9}-\frac{x^2}{16}=1$

17. $\frac{(x-4)^2}{4}-\frac{y^2}{12}=1$

18. $\frac{(y-1)^2}{9}-\frac{(x+3)^2}{16}=1$

19. $y=\pm\sqrt{\frac{2}{3}x^2+4}$

20. $y=\pm\sqrt{3x^2-6x}$

6.4 Summary of the Conic Sections

1. ellipse **2.** hyperbola **3.** hyperbola **4.** parabola

5. ellipse **6.** hyperbola **7.** ellipse **8.** ellipse

9. ellipse **10.** parabola **11.** hyperbola **12.** parabola

13. hyperbola **14.** parabola **15.** ellipse **16.** hyperbola

17. parabola **18.** hyperbola **19.** ellipse **20.** hyperbola

Write the first five terms of each sequence.

1. $a_n = n(n+1)$

2. $a_n = \frac{n^2 - 3}{3}$

3. $a_n = (-1)^n$

4. $a_n = 2^n - 1$

Find the first four terms of each sequence.

5. $a_1 = 4$
   $a_n = a_{n-1} + 3$ for $n > 1$

6. $a_1 = 5$
   $a_n = a_{n-1} - 3$ for $n > 1$

7. $a_1 = 2$
   $a_n = n \cdot a_{n-1}$ for $n > 1$

8. $a_1 = 81$
   $a_n = \frac{1}{3} a_{n-1}$ for $n > 1$

Evaluate each series.

9. $\sum_{i=1}^{5} i^2$

10. $\sum_{i=1}^{8} (7i - 6)$

11. $\sum_{i=1}^{5} \frac{i}{i+1}$

12. $\sum_{i=0}^{3} 3 \cdot 4^i$

13. $\sum_{i=1}^{10} (i - 2)$

14. $\sum_{i=1}^{3} \frac{1}{i^3}$

Use a graphing calculator to evaluate each series.

15. $\sum_{i=0}^{20} 100(1.06)^i$

16. $\sum_{i=0}^{10} 3 \cdot 2^i$

Write the terms for each series. Evaluate the sum given that
$x_1 = -4,\ x_2 = -2,\ x_3 = 0,\ x_4 = 2,$ and $x_5 = 4$.

17. $\sum_{i=1}^{5} \frac{2}{x_i + 1}$

18. $\sum_{i=1}^{5} \frac{3 + 4x_i}{5}$

Use the summation properties and rules to evaluate each series.

19. $\sum_{i=1}^{10} 3$

20. $\sum_{i=1}^{10} 6i$

21. $\sum_{i=1}^{5} 3\ i^2$

22. $\sum_{i=1}^{10} i^2 + 2i + 2$

Find the common difference d for each arithmetic sequence.

1. $3, 1, -1, -1, \ldots$
2. $6, 7.5, 9, 10.5, \ldots$
3. $0.25, 0.5, 0.75, 1, \ldots$
4. $x+2, 2x+4, 3x+6, 4x+8, \ldots$

Write the first five terms of each arithmetic sequence.

5. $a_1 = 7$; $d = 4$
6. $a_1 = 9$; $d = \frac{1}{3}$

Find a_5 and a_n for each arithmetic sequence.

7. $65, 80, 95, \ldots$
8. $-6, -1, 4, \ldots$
9. $a_1 = 2, a_4 = 8$
10. $a_2 = -5, a_4 = 5$

Evaluate S_8, the sum of the first eight terms of each arithmetic sequence.

11. $24, 16, 8, 0, \ldots$
12. $2, \frac{5}{2}, 3, \frac{7}{2}, \ldots$
13. $a_1 = 3, a_4 = 9$
14. $a_2 = 1, a_5 = -8$

Find a_1 and d for each arithmetic series.

15. $S_7 = -6, a_2 = 4$
16. $S_3 = -60, a_2 = -20$

Evaluate each sum.

17. $\sum_{i=1}^{5}(i+1)$
18. $\sum_{i=1}^{10}(2i+2)$
19. $\sum_{i=1}^{8}(2i-1)$
20. $\sum_{i=1}^{6}(5i-2)$

Find a_4 and a_n for each geometric sequence.

1. 8, 4, 2, ...

2. $-2,\ 6,\ -18,\ \ldots$

3. $a_1 = 1,\ r = 4$

4. $a_1 = 9,\ r = \frac{1}{3}$

Find a_1 and r for each geometric sequence.

5. $a_2 = 9,\ a_5 = \frac{1}{3}$

6. $a_3 = -6,\ a_9 = -\frac{3}{32}$

Use the formula for S_n to find the sum of the first five terms of each geometric sequence.

7. $20,\ 5,\ \frac{5}{4},\ \frac{5}{16},\ \ldots$

8. $\frac{2}{5},\ \frac{4}{5},\ \frac{8}{5},\ \ldots$

9. $a_1 = 2,\ r = 5$

10. $a_1 = 81,\ r = -\frac{1}{3}$

Find each sum.

11. $\sum_{i=1}^{5} \left(\frac{1}{5}\right)^i$

12. $\sum_{i=1}^{3} \left(-\frac{3}{10}\right)^i$

13. $\sum_{i=1}^{5} 2 \cdot 3^i$

14. $\sum_{i=1}^{3} 5 \cdot \left(\frac{3}{4}\right)^i$

Find r for each infinite geometric sequence. Identify any whose sum does not converge.

15. $2,\ -6,\ 18,\ \ldots$

16. $5,\ 1,\ \frac{1}{5},\ \frac{1}{25},\ \ldots$

Evaluate each sum that converges.

17. $\sum_{i=1}^{\infty} \left(\frac{1}{3}\right)^i$

18. $\sum_{i=1}^{\infty} 10(0.1)^{i-1}$

19. $\sum_{i=1}^{\infty} 5(1.07)^i$

20. $\sum_{i=1}^{\infty} 2\left(\frac{3}{5}\right)^i$

Evaluate each expression.

1. $\frac{9!}{5!4!}$

2. $\frac{5!}{3!2!}$

3. $\frac{6!}{2!4!}$

4. $\binom{6}{3}$

5. $\binom{4}{3}$

6. $\binom{7}{2}$

7. ${}_5C_5$

8. ${}_7C_3$

9. ${}_6C_4$

Write the binomial expansion for each expression.

10. $(x+4)^6$

11. $(x-2)^3$

12. $(1-x^2)^6$

13. $(x+y)^6$

14. $(x-2y)^4$

15. $(x+3y)^8$

16. $(3+2x)^4$

17. $(3x-1)^5$

Write the indicated term of each binomial expansion.

18. seventh term of $(2x+y)^8$

19. fifth term of $(x-y)^{10}$

20. tenth term of $(x^2-1)^{11}$

Assume that n is a positive integer. Use the method of mathematical induction to prove each statement.

1. $1^2+3^2+5^2+...+(2n-1)^2=\frac{1}{3}n(2n-1)(2n+1)$

2. $4+8+12+...+4n=2n(n+1)$

3. $5+7+9+...+(2n+3)=n(n+4)$

4. $1\cdot 2+2\cdot 3+3\cdot 4+...+n(n+1)=\frac{1}{3}n(n+1)(n+2)$

5. $1+5+9+...+(4n-3)=n(2n-1)$

6. $5+8+11+...+(3n+2)=\frac{1}{2}n(3n+7)$

7. $1\cdot 3+2\cdot 4+3\cdot 5+...+n(n+2)=\frac{1}{6}n(n+1)(2n+7)$

8. $2+5+8+...+(3n-1)=\frac{1}{2}n(3n+1)$

9. $2+2^2+2^3+...+2^n=2(2^n-1)$

10. $\cos(n\pi)=(-1)^n$

11. $\frac{1}{1\cdot 2\cdot 3}+\frac{1}{2\cdot 3\cdot 4}+\frac{1}{3\cdot 4\cdot 5}+...+\frac{1}{n(n+1)(n+2)}=\frac{n(n+3)}{4(n+1)(n+2)}$

12. $2^{n+3}<(n+3)!$

13. If $h\geq 0$, then $1+nh\leq(1+h)^n$.

14. $n<2^n$

15. $2\leq 2^n$

16. If $n\geq 7$, then $\left(\frac{4}{3}\right)^n>n$.

17. If $n\geq 2$, then $\frac{1}{\sqrt{1}}+\frac{1}{\sqrt{2}}+\frac{1}{\sqrt{3}}+...+\frac{1}{\sqrt{n}}>\sqrt{n}$.

18. If $n\geq 1$ and $0<a<b$, then $\left(\frac{a}{b}\right)^{n+1}<\left(\frac{a}{b}\right)^n$.

19. $\left(\frac{a}{b}\right)^n=\frac{a^n}{b^n}$

Evaluate each expression.

1. $C(6,4)$ 2. $P(4,2)$ 3. $C(5,3)$ 4. $P(8,3)$

5. ${}_{10}P_7$ 6. ${}_{20}C_4$ 7. ${}_{28}P_2$ 8. ${}_{15}C_6$

Use the fundamental principle of counting or permutations to solve each problem.

9. In how many ways can 2 cards be drawn in succession from a pack of 5 different cards marked !, @, #, $, and % ? Assume that the first card is not replaced before the second card is drawn.

10. In how many ways can 2 cards be drawn in succession from a pack of 5 different cards marked !, @, #, $, and % ? Assume that the first card is replaced before the second card is drawn.

11. A deli offers a value meal that consists of a sandwich, a bag of chips and a soft drink. Suppose there are 9 different kinds of sandwiches, 4 different kinds of chips, and 5 different kinds of soft drinks. How many different meals can be made?

12. In how many ways can 6 students be seated in a row of 6 seats?

13. How many different 2 digit numbers can be formed using the digits 2, 4, 6, and 8? Assume that the digits cannot be repeated.

14. How many different 3 digit numbers can be formed using the digits 2, 4, 6, and 8? Assume that the digits can be repeated.

15. In a club with 12 members, how many ways can a slate of 3 officers consisting of president, vice-president, and treasurer be chosen?

16. In how many ways can 9 players be assigned to the 4 positions on a team, assuming that any player can play any position?

17. Two nine sided dice, one pink and one white, are thrown. How many different outcomes are possible?

18. In a club with 9 women and 7 men members, how many 5-member committees can be chosen that have all men?

19. In a club with 9 women and 7 men members, how many 5-member committees can be chosen that have 4 women and 3 men?

20. If a penny, nickel, dime and quarter are tossed together, how many combinations can be formed?

1. Find the probability of picking either an ace or a king in a single drawing from a standard deck of cards.

2. An urn contains 3 red balls, 4 blue balls, 2 green balls, and 1 yellow ball. What is the probability of drawing a blue ball?

3. An urn contains 3 red balls, 4 blue balls, 2 green balls, and 1 yellow ball. If two balls are drawn in succession what is the probability of drawing a blue ball followed by a yellow ball?

4. Find the probability of rolling a 5 on a single roll of a die.

5. Find the probability of rolling a sum of 6 when 2 dice are rolled.

6. Find the probability of picking a red card from a deck of cards.

7. Find the probability of rolling a sum greater than 4 when 2 dice are rolled.

8. The probability of being left-handed is about .10. What are the odds against being left-handed?

9. What is the probability of throwing a six twice in succession with a single die?

10. Find the probability of obtaining no heads in tossing a pair of coins.

11. An urn contains 3 red balls, 4 blue balls, 2 green balls, and 1 yellow ball. What are the odds in favor of drawing a green ball?

12. An urn contains 3 red balls, 4 blue balls, 2 green balls, and 1 yellow ball. What are the odds against drawing a red ball?

13. An urn contains 3 red balls, 4 blue balls, 2 green balls, and 1 yellow ball. What is the probability of drawing 2 yellow balls in succession without replacing the first ball drawn?

14. What is the probability of picking an "i" from the word "Mississippi"?

15. A multiple-choice test question has five choices a – e. What is the probability of guessing the correct answer?

16. The digits 0 – 9 are written on a slip of paper and placed in a hat. What is the probability of selecting an odd digit if one slip is drawn from the hat?

17. The digits 0 – 9 are written on a slip of paper and placed in a hat. What is the probability of selecting a digit greater than 4 or an even digit if one slip is drawn from the hat?

7.1 Sequences and Series

1. 2, 6, 12, 20, 30

2. $-\frac{2}{3}, \frac{1}{3}, 2, \frac{13}{3}, \frac{22}{3}$

3. $-1, 1, -1, 1, -1$

4. 1, 3, 7, 15, 31

5. 4, 7, 10, 13,

6. $5, 2, -1, -4$

7. 2, 4, 12, 48

8. 81, 27, 9, 3

9. 55

10. 204

11. $\frac{71}{20}$

12. 255

13. 35

14. $\frac{251}{216}$

15. 3999.27

16. 6141

17. $\frac{2}{5}$

18. 3

19. 30

20. 330

21. 165

22. 515

7.2 Arithmetic Sequences and Series

1. -2

2. 1.5

3. 0.25

4. $x+2$

5. 7, 11, 15, 19, 23

6. $9, \frac{28}{3}, \frac{29}{3}, 10, \frac{31}{3}$

7. $a_n = 50 + 15n;\ a_5 = 125$

8. $a_n = -11 + 5n;\ a_5 = 14$

9. $a_n = 2n;\ a_5 = 10$

10. $a_n = -15 + 5n;\ a_5 = 10$

11. -32

12. 30

13. 80

14. -52

15. $a_1 = \frac{45}{7};\ d = -\frac{17}{7}$

16. $a_1;\ d = -20 - a_1$

17. 20

18. 130

19. 64

20. 93

7.3 Geometric Sequences and Series

1. $a_n = 8\left(\frac{1}{2}\right)^{n-1}$; $a_4 = 1$

2. $a_n = -2(-3)^{n-1}$; $a_4 = 54$

3. $a_n = 4^{n-1}$; $a_4 = 64$

4. $a_n = 9\left(\frac{1}{3}\right)^{n-1}$; $a_4 = \frac{1}{3}$

5. $a_1 = 27$; $r = \frac{1}{3}$

6. $a_1 = -24$; $r = \frac{1}{2}$

7. $\frac{1705}{64}$

8. $\frac{62}{5}$

9. 1562

10. 61

11. $\frac{781}{3125}$

12. $-\frac{237}{1000}$

13. 726

14. $\frac{555}{64}$

15. $r = -3$, does not converge

16. $r = \frac{1}{5}$

17. $\frac{1}{2}$

18. $\frac{100}{9}$

19. does not converge

20. 3

7.4 The Binomial Theorem

1. 126

2. 10

3. 15

4. 20

5. 4

6. 21

7. 1

8. 35

9. 15

10. $x^6 + 24x^5 + 240x^4 + 1280x^3 + 3840x^2 + 6144x + 4096$

11. $x^3 - 6x^2 + 12x - 8$

12. $x^{12} - 6x^{10} + 15x^8 - 20x^6 + 15x^4 - 6x^2 + 1$

13. $x^6 + 6x^5y + 15x^4y^2 + 20x^3y^3 + 15x^2y^4 + 6xy^5 + y^6$

14. $x^4 - 8x^3y + 24x^2y^2 - 32xy^3 + 16y^4$

15. $x^8 + 24x^7y + 252x^6y^2 + 1512x^5y^3 + 5670x^4y^4 + 13608x^3y^5 + 20412x^2y^6 + 17496xy^7 + 6561y^8$

16. $16x^4 + 96x^3 + 216x^2 + 216x + 81$

17. $243x^5 - 405x^4 + 270x^3 - 90x^2 + 15x - 1$

18. $112x^2y^6$

19. $210x^6y^4$

20. $-55x^4$

7.6 Counting Theory

1. 15 **2.** 12 **3.** 10 **4.** 336

5. 604,800 **6.** 4845 **7.** 756 **8.** 5005

9. 20 **10.** 25 **11.** 180 **12.** 720

13. 12 **14.** 64 **15.** 1320 **16.** 126

17. 81 **18.** 21 **19.** 4410 **20.** 16

7.7 Basics of Probability

1. $\frac{2}{13}$ **2.** $\frac{2}{5}$ **3.** $\frac{2}{45}$ **4.** $\frac{1}{6}$

5. $\frac{5}{36}$ **6.** $\frac{1}{2}$ **7.** $\frac{5}{6}$ **8.** $9:1$

9. $\frac{1}{36}$ **10.** $\frac{1}{4}$ **11.** $1:4$ **12.** $7:3$

13. 0 **14.** $\frac{4}{11}$ **15.** $\frac{1}{5}$ **16.** $\frac{1}{2}$

17. $\frac{4}{5}$